# *Mediaeval Lore from Bartholomew Anglicus*

*Robert Steele*

# Contents

# MEDIAEVAL LORE FROM BARTHOLOMEW ANGLICUS

BY

Robert Steele

"WHEN HOLY WERE THE HAUNTED FOREST BOUGHS,
HOLY THE AIR, THE WATER, AND THE FIRE."

KEATS.

# PREFACE

It is not long since the Middle Ages, of the literature of which this book gives us such curious examples, were supposed to be an unaccountable phenomenon accidentally thrust in betwixt the two periods of civilisation, the classical and the modern, and forming a period without growth or meaning--a period which began about the time of the decay of the Roman Empire, and ended suddenly, and more or less unaccountably, at the time of the Reformation. The society of this period was supposed to be lawless and chaotic; its ethics a mere conscious hypocrisy; its art gloomy and barbarous fanaticism only; its literature the formless jargon of savages; and as to its science, that side of human intelligence was supposed to be an invention of the time when the Middle Ages had been dead two hundred years.

The light which the researches of modern historians, archaeologists, bibliographers, and others, have let in on our view of the Middle Ages has dispersed the cloud of ignorance on this subject which was one of the natural defects of the qualities of the learned men and keen critics of the eighteenth and early part of the nineteenth centuries. The Middle-class or Whig theory of life is failing us in all branches of human intelligence. Ethics, Politics, Art, and Literature are more than beginning to be regarded from a wider point of view than that from which our fathers and grandfathers could see them.

For many years there has been a growing reaction against the dull "grey" narrowness of the eighteenth century, which looked on Europe during the last thousand years as but a riotous, hopeless, and stupid prison. It is true that it was on the side of Art alone that this enlightenment began, and that even on that side it progressed slowly enough at first--e.g. Sir Walter Scott feels himself obliged, as in the *Antiquary*, to apologize to pedantry for his instinctive love of Gothic architecture. And no less true is it that follies enough were mingled with the really useful and healthful birth of romanticism in Art and Literature. But at last the study of facts by men who were neither artistic nor sentimental came to the help of that first glimmer of instinct, and gradually something like a true insight into the life of the Middle Ages was gained; and we see that the world of Europe was no more running round in a circle then than now, but was developing, sometimes with stupendous speed, into something as different from itself as the age which succeeds this will be different from that wherein we live. The men of those times are no longer puzzles to us; we can understand their aspirations, and sympathise with their lives, while at the same time we have no wish (not to say hope) to put back the clock, and start from the position which they held. For, indeed, it is characteristic of the times in which we live, that whereas in the beginning of the romantic reaction, its supporters were for the most part mere ***laudatores temporis acti***, at the present time those who take pleasure in studying the life of the Middle Ages are more commonly to be found in the ranks of those who are pledged to the forward movement of modern life; while those who are vainly striving to stem the progress of the world are as careless of the past as they are fearful of the future. In short, history, the new sense of modern times, the great compensation for the losses of the centuries, is now teaching us worthily, and making us feel that the past is not dead, but is living in us, and will be alive in the future which we are now helping to make.

To my mind, therefore, no excuse is needful for the attempt made in the following pages to familiarise the reading public with what was once a famous knowledge-book of the Middle Ages. But the reader, before he can enjoy it, must cast away the exploded theory of the invincible and wilful ignorance of the days when it was written; the people of that time were eagerly desirous for knowledge, and their teachers were mostly single-hearted and intelligent men, of a diligence and laboriousness almost past belief. The "Properties of Things" of Bartholomew the Englishman is but one of the huge encyclopaedias written in the early Middle Age for the instruction of those who wished to learn, and the reputation of it and its fellows shows how much the science of the day was appreciated by the public at large, how many there were who wished to learn. Even apart from its interest as showing the tendency of men's minds in days when Science did actually tell them "fairy tales," the book is a delightful one in its English garb; for the language is as simple as if the author were speaking by word of mouth, and at the same time is pleasant, and not lacking a certain quaint floweriness, which makes it all the easier to retain the subject-matter of the book.

Altogether, this introduction to the study of the Mediaeval Encyclopaedia, and the insight which such works give us into the thought of the past and its desire for knowledge, make a book at once agreeable and useful; and I repeat that it is a hopeful sign of the times when students of science find themselves drawn towards the historical aspect of the world of men, and show that their minds have been enlarged, and not narrowed, by their special studies--a defect which was too apt to mar the qualities of the seekers into natural facts in what must now, I would hope, be called the just-passed epoch of intelligence dominated by Whig politics, and the self-sufficiency of empirical science.

WILLIAM MORRIS.

# INTRODUCTION

THE BOOK AND ITS OBJECT.--The book which we offer to the public of to-day is drawn from one of the most widely read books of mediaeval times. Written by an English Franciscan, Bartholomew, in the middle of the thirteenth century, probably before 1260, it speedily travelled over Europe. It was translated into French by order of Charles V. (1364-81) in 1372, into Spanish, into Dutch, and into English in 1397. Its popularity, almost unexampled, is explained by the scope of the work, as stated in the translator's prologue (p. 9). It was written to explain the allusions to natural objects met with in the Scriptures or in the Gloss. It was, in fact, an account of the properties of things in general; an encyclopaedia of similes for the benefit of the village preaching friar, written for men without deep--sometimes without any-- learning. Assuming no previous information, and giving a fairly clear statement of the state of the knowledge of the time, the book was readily welcomed by the class for which it was designed, and by the small nucleus of an educated class which was slowly forming. Its popularity remained in full vigour after the invention of printing, no less than ten editions being published in the fifteenth century of the Latin copy alone, with four French translations, a Dutch, a Spanish, and an English one.

The first years of the modern commercial system gave its death-blow to the popularity of this characteristically mediaeval work, and though an effort was made in 1582 to revive it,  the attempt was unsuccessful--quite naturally so, since the book was written for men desirous to hear of the wonders of strange lands, and did not give an accurate account of anything. The man who bought cinnamon at Stourbridge Fair in 1380 would have felt poorer if any one had told him that it was not shot from the phoenix' nest with leaden arrows, while the merchant of 1580 wished to know where it was grown, and how much he would pay a pound for it if

he bought it at first hand. Any attempt to reconcile these frames of mind was fore-
doomed to failure.

THE INTEREST OF BARTHOLOMEW'S WORK.--The interest of Bartho-
lomew's work to modern readers is twofold: it has its value as literature pure and
simple, and it is one of the most important of the documents by the help of which
we rebuild for ourselves the fabric of mediaeval life. The charm of its style lies in its
simple forcible language, and its simplicity suits its matter well. On the one hand,
we cannot forget it is a translation, but the translation, on the other hand, is from
the mediaeval Latin of an Englishman into English.

One of the greatest difficulties in the way of a student is to place himself in
the mental attitude of a man of the Middle Ages towards nature; yet only by so doing can he appreciate the solutions that the philosophers of the time offered of the
problems of nature. Our author affords perhaps the simplest way of learning what
Chaucer and perhaps Shakespeare knew and believed of their surroundings--earth,
air, and sea. The plan on which his work was constructed led Bartholomew in order
over the universe from God and the angels--through fire, water, air, to earth and all
that therein is. We thus obtain a succinct account of the popular mediaeval theories
in Astronomy, Physiology, Physics, Chemistry, Geography, and Natural History,
all but unattainable otherwise. The aim of our chapter on Science has been to give
sufficient extracts to mark the theories on which mediaeval Science was based, the
methods of its reasoning, and the results at which it arrived. The chapter on Medicine gives some account of the popular cures and notions of the day, and that on
Geography resumes the traditions current on foreign lands, at a time when Ireland
was at a greater distance than Rome, and less known than Syria.

In the chapter on Mediaeval Society we have not perhaps the daily life of the
Middle Ages, but at least the ideal set before them by their pastors and masters--an
ideal in direct relationship with the everyday facts of their life. The lord, the servant, the husband, the wife, and the child, here find their picture. Some information, too, can be obtained about the daily life of the time from the chapter on the
Natural History of Plants, which gives incidentally their food-stuffs.

It is in the History of Animals that the student of literature will find the richest
mine of allusions. The list of similes in Shakespeare explained by our author would
fill a volume like this itself. Other writers, again, simply "lift" the book wholesale.

Chester and Du Bartas write page after page of rhyme, all but versified direct from Bartholomew. Jonson and Spenser, Marlowe and Massinger, make ample use of him. Lyly and Drayton owe him a heavy debt. Considerations of space forbid their insertion, but for every extract made here, the Editor has collected several passages from first-class authors with a view to illustrating the immense importance of this book to Elizabethan literature. It was not without reason that Ireland chose justified, when making a selection of passages from the work for modern readers, in altering his text to this extent--and this only: he has modernised the spelling, and in the case of entirely obsolete grammatical forms he has substituted modern ones (e.g. "its" for "his"). In the case of an utterly dead word he has followed the course of substituting a word from the same root, when one exists; and when none could be found, he has left it unchanged in the text. Accordingly a short glossary has been added, which includes, too, many words which we may hope are not dead, but sleeping. In very few cases has a word been inserted, and in those it is marked by italics.

Perhaps we may be allowed to say a word in defence of the principle of modernising our earliest literature. Early English poetry is, in general (with some striking exceptions), incapable of being written in the spelling of our days without losing all of that which makes it verse; but there can be no reason, when dealing with the masterpieces of our Early English prose, for maintaining obsolete forms of spelling and grammar which hamper the passage of thought from mind to mind across the centuries. Editors of Shakespeare and the Bible for general use have long assumed the privilege of altering the spelling, and except on the principle that earlier works are more important, or are only to be read by people who have had the leisure and inclination to familiarise their eyes with the peculiarities of Middle English, there can be no reason for stopping there, or a century earlier. At some point, of course, the number of obsolete words becomes so great that the text cannot be read without a dictionary: then the limit has been reached. But Caxton, Trevisa, and many others are well within it, and it is good to remove all obstacles which prevent the ordinary reader from feeling the continuity of his mother tongue.

THE AUTHOR.--The facts known of our author's life have been summarised by Miss Toulmin Smith in her article in the ***Dictionary of National Biography***. In the sixteenth century he was generally believed to date from about 1360, and to have belonged to the Glanvilles--an honourable Suffolk family in the Middle Ages;

but there seems to be no authority whatever for the statement. We first hear of him in a letter from the provincial of the Franciscans of Saxony to the provincial of France, asking that Bartholomew Anglicus and another friar should be sent to assist him in his newly-created province. Next year (1231) a MS. chronicle reports that two were sent, and that Bartholomew Anglicus was appointed teacher of holy theology to the brethren in the province. We learn from Salimbene, who wrote the Chronicles of Parma (1283), that he had been a professor of theology in the University of Paris, where he had lectured on the whole Bible. The subject in treating of which he is referred to was an elephant belonging to the Emperor; and Salimbene quotes a passage on the elephant from his *De Proprietatibus Rerum*. What may be a quotation from the *De Proprietatibus* can be found in Roger Bacon's *Opus Tertium* (1267).

THE DATE OF THE WORK.--The date of the work seems fairly easy to fix. It cannot, as we have above seen, be later than 1267, and Amable Jourdain fixes it before 1260 by the fact that the particular translations of Aristotle from which Bartholomew quotes (Latin through the Arabic), went almost universally out of use by 1260. On the other hand, quotations are made from Albertus Magnus, who was in Paris in 1248. And that it was written near this year is evident from the fact that no quotations are made from Vincent of Beauvais, Thomas Aquinas, Roger Bacon, or Egidius Colonna, all of whom were in Paris during the second half of the thirteenth century. The earliest known MS. is in the Ashmole Collection, and was written in 1296. Two French MSS. are dated 1297 and 1329 respectively.

As we said in the beginning of this chapter, the work had an immediate and lasting success. Bartholomew Anglicus became known as "Magister de Proprietatibus Rerum," and his book was on the list of those which students could borrow from the University chest. It is probable that much of this popularity was due to the fact that he was a teacher for many years of the Grey Friars, and that these, the most popular and the most human preachers of the day, carried his book and his stories with them wherever they went.

SOURCES.--The chief sources of our author's inspiration are notable. He relies on St. Dionysius the Areopagite for heaven and the angels, Aristotle for Physics and Natural History, Pliny's Natural History, Isidore of Seville's Etymology, Albumazar, Al Faragus, and other Arab writers for Astronomy, Constantinus Afer's Pantegna for Medical Science, and Physiologus, the Bestiarium, and the Lapidarium for the properties of gems, animals, etc. Besides these he quotes many other writers (a list of whom is given in an appendix) little known to modern readers.

THE TRANSLATION AND PRINCIPLES OF SELECTION.--The translation from which we quote was made for Sir Thomas lord of Berkeley in 1397 by John Trevisa, his chaplain. We owe this good Englishman something for the works in English prose he called into existence--some not yet printed; may we not see in him another proof of what we owe to Chaucer--a language stamped with the seal of a great poet, henceforth sufficient for the people who speak it, ample for the expression of their thoughts or needs?

In selecting from such a book, the principles which have guided the editor are these: To the general reader he desires to offer a fair representation of the work of Bartholomew Anglicus, preserving the language and style. To be fair, the work must be sometimes dull--in the whole book there are many very dull passages. He has desired to select passages of interest for their quaint language, and their views of things, often for their very misrepresentations of matters of common knowledge to-day, and for their bearing upon the literature of the country. The student of literature and science will find in it the materials in which the history of their growth is read. In conclusion, the editor ventures to hope that the work will not be unwelcome to the numerous and growing class who love English for its own sake as the noblest tongue on earth, and who desire not to forget the rock from which it was hewn, and the pit from which it was digged.

Our first selection will naturally be the translator's prologue in the very shortened form of Berthelet. The present editor's work is, to avoid confusion, printed in small type throughout.

# THE PROLOGUE OF THE TRANSLATOR

True it is that after the noble and expert doctrine of wise and well-learned Philosophers, left and remaining with us in writing, we know that the properties of things follow and ensue their substance. Herefore it is that after the order and the distinction of substances, the order and the distinction of the properties of things shall be and ensue. Of the which things this work of all the books ensuing, by the grace, help, and assistance of all mighty God is compiled and made.

Marvel not, ye witty and eloquent readers, that I, thin of wit and void of cunning, have translated this book from Latin into our vulgar language, as a thing profitable to me, and peradventure to many other, which understand not Latin, nor have not the knowledge of the properties of things, which things be approved by the books of great and cunning clerks, and by the experience of most witty and noble Philosophers. All these properties of things be full necessary and of great value to them that will be desirous to understand the obscurities, or darkness of holy scriptures: which be given unto us under figures, under parables and semblance, or likelihoods of things natural and artificial. Saint Denys, that great Philosopher and solemn clerk, in his book named the heavenly hierarchies of angels, testifieth and witnesseth the same, saying in this manner:--What so ever any man will conject, feign, imagine, suppose, or say: it is a thing impossible that the light of the heavenly divine clearness, covered and closed in the deity, or in the godhead, should shine upon us, if it were not by the diversities of holy covertures. Also it is not possible, that our wit or intendment might ascend unto the contemplation of the heavenly hierarchies immaterial, if our wit be not led by some material thing, as a man is led by the hand: so by these forms visible, our wit may be led to the consideration of the greatness or magnitude of the most excellent beauteous clarity, divine and invisible. Reciteth this also the blessed apostle Paul in his epistles, saying that by these things visible, which be made and be visible, man may see and know by his inward sight intellectual, the divine celestial and godly things, which be invisible to this our natural sight. Devout doctors of Theology or divinity, for this consideration prudently and wisely read and use natural philosophy and moral, and poets in their fictions and feigned informations, unto this fine and end, so that by the likelihood or similitude of things visible our wit or our understanding spiritually, by clear and crafty utterance of words, may be so well ordered and uttered: that these things corporeal may be coupled with things spiritual, and that these things visible may be conjoined with things Invisible. Excited by these causes to the edifying of the people contained in our Christian faith of almighty Christ Jesus, whose majesty divine is incomprehensible: and of whom to speak it becometh no man, but with great excellent worship and honour, and with an inward dreadful fear. Loth to offend, I purpose to say somewhat under the correction of excellent learned doctors and wise men: what every creature reasonable ought to believe in this our blessed Christian faith.

ENDETH THE PROLOGUE

# I
# MEDIAEVAL SCIENCE

The following selections will give an idea of the natural science of the Middle Ages. In introducing them, the Editor will attempt to give some connected account of them to show that though their study seems to involve a few difficulties, their explanation is simple, and will not make too great a demand on the reader's patience.

From the earliest times men have asked themselves two questions about nature: "Why?" and "How?" Mediaeval science concerned itself with the former; modern science thinks it has learnt that no answer to that question can be given it, and concerns itself with the latter. It thus happens that the more one becomes in sympathy with the thought of our time, the less one can interest one's self in the work of the past, distinguished as it is by its disregard of all we think important, and by its striving for an unattainable goal.

It is, however, necessary, if we would enjoy Chaucer, Dante, and Shakespeare, to obtain some notion of that system of the universe from which they drew so many of their analogies. The symbolism of Dante appears to us unnaturally strained until we know that the science of his day saw everything as symbolic.

And how could we appreciate the strength of Chaucer's metaphor:

"O firste moving cruel firmament,
With thy diurnal swegh that croudest ay,
And hurtlest all from Est til Occident,
That naturally wold hold another way,"

without some knowledge of the astronomy of his day?

Our first extracts explain themselves. They deal with the mystery of the constitution of substances, as fascinating to us as to the early Greeks, and begin with definitions of matter and form.

The principal design of early philosophers in physics was to explain how everything was generated, and to trace the different states through which things pass until they become perfect. They observed that as a thing is not generated out of any other indifferently--for example, that marble is not capable of making flesh, all bodies cannot be compounded of principles alone, connected in a simple way, but imagined they could be made up of a few simple compounds. These ultimate compounds, if we may so express it, were their elements. The number of elements was variously estimated, but was generally taken as four--a number arrived at rather from the consideration of the sensations bodies awaken in us, than from the study of bodies themselves. Aristotle gives us the train of thought by which the number is reached. He considers the qualities observed by the senses, classifying them as Heat, Cold, Dryness or Hardness, and Moistness or Capability of becoming liquid. These may partially co-exist, two at a time, in the same substance. There are thus four possible combinations, Cold and dry, Cold and moist, Hot and dry, Hot and moist. He then names these from their prototypes Earth, Water, Fire, and Air, distinguishing these elements from the actual Earth, etc., of everyday life.

The habit of extending analogies beyond their legitimate application was a source of confusion in the early ages of science. Most of the superstitions of primitive religion, of astrology, and of alchemy, arose from this source. A good example is the extension of the metaphor in the words *generation* and *corruption*: words in constant use in scientific works until the nineteenth century began. Generation is the production of a substance that before was not, and corruption is the destruction of a substance, by its ceasing to be what it was before. Thus, fire is generated, and wood is corrupted, when the latter is burnt. But the implicit metaphor in the use of the terms likens substances to the human body, their production and destruction implies liability to disease, and thus prepares the way for the notion of the elixir, which is first a potion giving long life, and curing bodily ailments, and only after some time a remedy for diseased metals--the philosopher's stone.

It will be seen that the theory of the mediaeval alchemist was that matter is an entity filling all space, on which in different places different forms were impressed.

The elements were a preliminary grouping of these, and might be present--two, three, or four at a time--in any substance. No attempt was ever made to separate these elements by scientific men, just as no attempt is ever made to isolate the ether of the physical speculations of to-day. The theory of modern physicists, with its ether and vortices, answers almost exactly to the matter and form of the ancients, the nature of the vortices conditioning matter.

The extracts from Book XI. bring us to another class of substances. All compound bodies are classified as imperfect or perfect. Imperfect compounds, or meteors, to some extent resemble elements. They are fiery, as the rainbow, or watery, as dew. Our extract on the rainbow is somewhat typical of the faults of ancient science. A note is taken of a rare occurrence--a lunar rainbow; but in describing the common one, an error of the most palpable kind is made. The placing of blue as the middle and green as the lowest colour is obviously wrong, and is inexplicable if we did not know how facts were cut square with theories in old days.

In the next extract Bartholomew's account of the spirits animating man is quoted at length. It gives us the mediaeval theory as to the means by which life, motion, and knowledge were shown in the body. Every reader of Shakespeare or Chaucer becomes familiar with the vital, animal, and natural spirits. They were supposed to communicate with all parts of the body by means of the arteries or wosen, "the nimble spirits in their arteries," and the sinews or nerves. The word sinew, by the way, is exactly equal to our word nerve, and ayenward, as our author would say. Hamlet, when he bursts from his friends, explains his vigour by the rush of the spirit into the arteries, which makes

"Each petty artery of this body
As hardy as the Nemean lion's nerve."

The natural spirit is generated in the liver, the seat of digestion, "there where our nourishment is administered"; it then passes to the heart, and manifests itself as the spirit of life; from thence it passes to the brain, where it is the animal spirit-- "spirit animate" Rossetti calls it--dwelling in the brain.

In the brain there are three ventricles or chambers, the *foremost* being the "cell fantastike" of the "Knight's Tale," the second the logistic, and the third the chamber of memory, where "memory, the warder of the brain," keeps watch over the passage of the spirit into the "sinews" of moving. Into the foremost cell come all the perceptions of sight, hearing, etc., and thus we have the opportunity for

"Fantasy,
That plays upon our eyesight,"

to freak it on us. The pedant, Holofernes, in **Love's Labour's Lost,** characteristically puts the origin of his good things in the ventricle of memory.

As a specimen of the physical science of the time the Editor gives extracts from the chapter on light.

The introduction of extracts enough to give some idea of the mediaeval astronomy would have made such large demands on the patience of the reader that the Editor has decided with some regret to omit them altogether. The universe is considered to be a sphere, whose centre is the earth and whose circumference revolved about two fixed points. Our author does not decide the nice point in dispute between the philosophers and the theologians, the former holding that there is only one, the latter insisting on seven heavens-the fairy, ethereal, olympian, fiery, firmament, watery, and empyrean.

The firmament, that

"Majestical roof, fretted with golden fire,"

is the part of heaven in which the planets move. It carries them round with it; it governs the tides; it stood with men for the type of irresistible regularity. Each of the planets naturally has a motion of its own, contrary in direction to that of the firmament, which was from east to west. All the fixed stars move in circles whose centre is the centre of the universe, but the courses of the planets (among which the moon is reckoned) depend on other circles, called eccentric, since their centre is elsewhere. Either the centre or the circumference of the circle in which the planet really moves is applied to the circumference of the eccentric circle, and in this way all the movements of the planets are fully explained. Our author is sorely puzzled to account for the existence of the watery heavens above the fiery, they being cold and moist, but is sure from scriptural reasons that they are there, and ventures the hypothesis that their presence may account for the sluggish and evil properties of Saturn, the planet whose circle is nearest them.

Having considered the simpler substances, those composed of pure elemental forms, and those resembling them--the meteors--we turn to the perfect compounds, those which have assumed substantial forms, as metals, stones, etc. Our author retains the Aristotelian classification--earthy, and those of other origin, as beasts, roots, and trees. Earths may be metals or fossils; metals being defined as hard bodies, generated in the earth or in its veins, which can be beaten out by a hammer, and softened or liquefied by heat; while fossils include all other inanimate objects.

A large number of extracts have been made from this part of the subject, because the book gives the position of positive, as distinguished from speculative, Alchemy at the time. It is the Editor's desire to show that at this period there was a system of theory based on the practical knowledge of the day.

Chemistry took its rise as a science about four hundred years before our era. In the fragments of two of the four books of Democritus we have probably the earliest treatise on chemical matters we are ever likely to get hold of. Whether it is the work of Democritus or of a much later writer is uncertain. But merely taking it as a representative work of the early stage of chemistry, we remark that the receipts are practicable, and some of them, little modified, are in use to-day in goldsmith's shops. The fragments remaining to us are on the manufacture of gold and silver, and one receipt for dyeing purple. In this state of the science the collection of facts is the chief point, and no purely chemical theory seems to have been formed. Tradition, confirmed by the latest researches, associates this stage with Egypt.

The second stage in the history of Chemistry--the birth of Alchemy in the Western World--occurred when the Egyptian practical receipts, the neo-Greek philosophies, and the Chinese dreams of an "elixir vitae" were fused into one by the Arab and Syriac writers. Its period of activity ranges from the seventh to the tenth centuries. Little is really known about it, or can be, until the Arabic texts, which are abundant in Europe, are translated and classified both from the scholar's and the chemist's standpoint. Many works were translated into Latin about the end of the tenth century, such as the spurious fourth book of the *Meteorics of Aristotle*, the treatises of the *Turta Philosophorum*, *Artis Auriferae*, etc., which formed the starting-point of European speculation. The theoretical chemistry of our author is derived from them.

The third stage of chemistry begins with the fourteenth and ends with the sixteenth century. It is characterized by an immense growth of theory, a fertile imagination, and untiring industry. It reached its height in England about 1440, and is represented by the reputed works of Lully (vixit circ. 1300), which first appeared about this date. In this period practical alchemy is on its trial.

The fourth stage begins with Boyle, and closes with the eighteenth century. Still under the dominion of theoretical alchemy, practical alchemy was rejected by it, and its interest was concentrated on the collection of facts. It led up to modern chemistry, which begins with Lavoisier, and the introduction of the balance in the study of chemical change.

Chemical theory, then, in our author's time stood somewhat thus. Metals as regarded their elemental composition were considered to partake of the nature of earth, water, and air, in various proportions. Fossils, or those things generated in the earth which were not metals, were again subdivided into two classes--those which liquefy on being heated, as sulphur, nitre, etc., and those which do not. The metals were considered to be composed of sulphur and mercury. These substances are themselves compounds, but they act as elements in the composition of metals. Sulphur represented their combustible aspect, and also that which gave them their solid form; while mercury was that to which their weight and powers of becoming fluid were due.

This theory was due to two main facts. Most ores of metals, especially of copper and lead, contain much sulphur, which can be either obtained pure from them, or be recognised by its smell when burning. This gave rise to the sulphur theory, while the presence of mercury was inferred doubtless from the resemblance of the more commonly molten metals, silver, tin, and lead, to quicksilver. The properties of each metal were then put down to the presence of these substances. The list of seven metals is that of the most ancient times--gold, electrum, silver, copper, tin, lead, iron; but it is clearly recognised that electrum is an alloy of gold and silver.

Most of the facts in this book are derived from Pliny through Isidore, but, that the theory is Arab in origin, one fact alone would convince us. A consideration of the composition of the metals shows us that tin is nearest in properties of all metals to the precious ones, but tin is precisely the metal chosen by Arab alchemists as a starting-point in the Chrysopoeia.

Beside their scientific interest these passages have supplied many analogies. When Troilus is piling up his lover's oaths to Cressida, his final words are:

"As iron to adamant, as earth to centre;"

our chapter on the adamant supplies the origin of this allusion in part, astronomy gives the other. Diamonds are still, unfortunately, the precious stones of reconciliation and of love our author bespeaks them. The editor has not lengthened the chapter by extracts giving the occult properties of gems, and has contented himself by quoting from the chapter on glass a new simile and an old story.

Matter and form are principles of all bodily things; and privation of matter and form is naught else but destruction of all things. And the more subtle and high matter is in kind, the more able it is to receive form and shape. And the more thick and earthly it is, the more feeble is it to receive impression, printing of forms and of shapes. And matter is principle and beginning of distinction, and of diversity, and of multiplying, and of things that are gendered. For the thing that gendereth and the thing that is gendered are not diverse but touching matter. And therefore where a thing is gendered without matter, the thing that gendereth, and the thing that is gendered, are all one in substance and in kind: as it fareth of the persons in the Trinity. Of form is diversity, by the which one thing is diverse from another, and some form is essential, and some accidental. Essential form is that which cometh into matter, and maketh it perfect; and accordeth therewith to the perfection of some thing. And when form is had, then the thing hath its being, and when form is destroyed nothing of the substance of the thing is found. And form accidental is not the perfection of things, nor giveth them being. But each form accidental needeth a form substantial. And each form is more simple and more actual and noble than matter. And so the form asketh that shall be printed in the matter, the matter ought to be disposed and also arrayed. For if fire shall be made of matter of earth, it needeth that the matter of earth be made subtle and pured and more simple. Form maketh matter known. Matter is cause that we see things that are made, and so nothing is more common and general than matter. And natheless nothing is more unknown than is matter; for matter is never seen without form, nor form may not be seen in deed, but joined to matter.

Elements are simple, and the least particles of a body that is compound. And it is called least touching us, for it is not perceived by wits of feeling. For it is the least part and last in undoing of the body, as it is first in composition. And is called simple, not for an element is simple without any composition, but for it hath no parts that compound it, that be diverse in kind and in number as some medlied bodies have: as it fareth in metals of the which some parts be diverse; for some part is air, and some is earth. But each part of fire is fire, and so of others. Elements are four, and so there are four qualities of elements, of the which every body is composed and made as of matter. The four elements are Earth, Water, Fire, and Air, of the which each hath his proper qualities. Four be called the first and principal qualities, that is, hot, cold, dry, and moist: they are called the first qualities because they slide first from the elements into the things that be made of elements. Two of these qualities are called Active--heat and coldness. The others are dry and wetness and are called Passive.

The Rainbow is impression gendered in an hollow cloud and dewy, disposed to rain in endless many gutters, as it were shining in a mirror, and is shapen as a bow, and sheweth divers colours, and is gendered by the beams of the sun or of the moon. And is but seldom gendered by beams of the moon, no more but twice in fifty years, as Aristotle saith. In the rainbow by cause of its clearness be seen divers forms, kinds, and shapes that be contrary. Therefore the bow seemeth coloured, for, as Bede saith, it taketh colour of the four elements. For therein, as it were in any mirror, shineth figures and shapes and kinds of elements. For of fire he taketh red colour in the overmost part, and of earth green in the nethermost, and of the air a manner of brown colour, and of water somedeal blue in the middle. And first is red colour, that cometh out of a light beam, that touches the outer part of the roundness of the cloud: then is a middle colour somedeal blue, as the quality asketh, that hath mastery in the vapour, that is in the middle of the cloud. Then the nethermost seemeth a green colour in the nether part of a cloud; there the vapour is more earthly. And these colours are more principal than others.

As Beda saith, and the master of stories, forty years tofore the doom, the rainbow shall not be seen, and that shall be token of drying, and of default of elements.

And though dew be a manner of airy substance, and most subtle outward, natheless in a wonder manner it is strong in working and virtue. For it besprinkleth the earth, and maketh it plenteous, and maketh flour, pith, and marrow increase in corn and grains: and fatteth and bringeth forth broad oysters and other shell fish in the sea, and namely dew of spring time. For by night in spring time oysters open themselves against dew, and receive dew that cometh in between the two shells, and hold and keep it; and that dew so holden and kept feedeth the flesh, and maketh it fat; and by its incorporation with the inner parts of the fish breedeth a full precious gem, a stone that is called Margarita. Also the birds of ravens, while they are whitish in feathers, ere they are black, dew feedeth and sustaineth them, as Gregory saith.

Fumosities that are drawn out of the waters and off the earth by strength of heat of heaven are drawn to the nethermost part of the middle space of the air, and there by coldness of the place they are made thick, and then by heat dissolving and departing the moisture thereof and not wasting all, these fumosities are resolved and fall and turn into rain and showers.

If rain be temperate in quality and quantity, and agreeable to the time, it is profitable to infinite things. For rain maketh the land to bear fruit, and joineth it together, if there be many chines therein, and assuageth and tempereth strength of heat, and cleareth the air, and ceaseth and stinteth winds, and fatteth fish, and helpeth and comforteth dry complexion. And if rain be evil and distemperate in its qualities, and discording to place and time, it is grievous and noyful to many things. For it maketh deepness and uncleanness and slipperiness in ways and in paths, and bringeth forth much unprofitable herbs and grass, and corrupteth and destroyeth fruit and seeds, and quencheth in seeds the natural heat, and maketh darkness and thickness in the air, and taketh from us the sun beams, and gathereth mist and clouds, and letteth the work of labouring men, and tarrieth and letteth ripening of corn and of fruits, and exciteth rheum and running flux, and increaseth and strengtheneth all moist ills, and is cause of hunger and of famine, and of corruption and murrain of beasts and sheep; for corrupt showers do corrupt the grass and herbs of pasture, whereof cometh needful corruption of beasts.

Of impressions that are gendered in the air of double vapour, the first is thunder, the which impression is gendered in watery substance of a cloud. For moving and shaking hither and thither of hot vapour and dry, that fleeth its contrary, is beset and constrained in every side, and smit into itself, and is thereby set on fire and on flame, and quencheth itself at last in the cloud, as Aristotle saith. When a storm of full strong winds cometh in to the clouds, and the whirling wind and the storm increaseth, and seeketh out passage: it cleaveth and breaketh the cloud, and falleth out with a great rese and strong, and all to breaketh the parts of the cloud, and so it cometh to the ears of men and of beasts with horrible and dreadful break= ing and noise. And that is no wonder: for though a bladder be light, yet it maketh great noise and sound, if it be strongly blown, and afterward violently broken. And with the thunder cometh lightning, but lightning is sooner seen, for it is clear and bright; and thunder cometh later to our ears, for the wit of sight is more subtle than the wit of hearing. As a man seeth sooner the stroke of a man that heweth a tree, than he heareth the noise of the stroke.

The lightning which is called Clarum is of a wonderful kind, for it catcheth and draweth up wine out of the tuns, and toucheth not the vessel, and melteth gold and silver in purses, and melteth not the purse.

As wits and virtues are needed to the ruling of kind, so to the perfection thereof needeth needly some spirits, by whose benefit and continual moving, both wits and virtues in beasts are ruled to work and do their deeds. As we speak here of a spirit, a spirit is called a certain substance, subtle and airy, that stirreth and exciteth the vir= tues of the body to their doings and works. A spirit is a subtle body, by the strength of heat gendered, and in man's body giving life by the veins of the body, and by the veins and pulses giveth to beasts, breath, life, and pulses, and working, wilful moving, and wit by means of sinews and muscles in bodies that have souls. Physi= cians say that this spirit is gendered in this manner wise. Whiles by heat working in the blood, in the liver is caused strong boiling and seething, and thereof cometh a smoke, the which is pured, and made subtle of the veins of the liver. And tur= neth into a subtle spiritual substance and airly kind, and that is called the natural spirit. For kindly by the might thereof it maketh the blood subtle. And by lightness thereof it moveth the blood and sendeth it about into all the limbs. And this same spirit turneth to heartward by certain veins. And there by moving and smiting together of the parts of the heart, the spirit is more pured, and turned into a more subtle kind. And then it is called of physicians the vital spirit: because that from the heart, by the wosen, and veins, and small ways, it spreadeth itself into all the limbs of the body, and increaseth the virtues spiritual, and ruleth and keepeth the works thereof. For out of a den of the left side of the heart cometh an artery vein, and in his moving is departed into two branches: the one thereof goeth downward, and spreadeth in many boughs, and sprays, by means of which the vital spirit is brought to give the life to all the nether limbs of the body. The other bough goeth upward, and is again departed in three branches. The right bough thereof goeth to the right arm, and the left bough to the left arm equally, and spreadeth in divers sprays. And so the vital spirit is spread into all the body and worketh in the artery veins the pulses of life. The middle bough extendeth itself to the brain, and other higher parts and giveth life, and spreadeth the vital spirit in all the parts about. The same spirit piercing and passing forth to the dens of the brain, is there more directed and made subtle, and is changed into the animal spirit, which is more subtle than the other. And so this animal spirit is gendered in the foremost den of the brain, and is somewhat spread into the limbs of feeling. But yet nevertheless some part thereof abideth in the aforesaid dens, that common sense, the common wit, and the virtue imaginative may be made perfect. Then he passeth forth into the middle den that is called Logistic, to make the intellect and understanding perfect. And when he hath enformed the intellect, then he passeth forth to the den of memory, and bearing with him the prints of likeness, which are made in those other dens, he layeth them up in the chamber of memory. From the hindermost parts of the brain he pierceth and passeth by the marrow of the ridge bone, and cometh to the sinews of moving, that so wilful moving may be engendered, in all the parts of the nether body. Then one and the same spirit is named by divers names. For by working in the liver it is called the natural spirit, in the heart the vital spirit, and in the head, the animal spirit. We may not believe that this spirit is man's reasonable soul, but more soothly, as saith Austin, the car therof and proper instrument. For by means of such a spirit the soul is joined to the body: and without the service of such a spirit, no act the soul may perfectly exercise in the body. And therefore if these spirits be impaired, or let of their working in any work, the accord of the body and soul is resolved, the reasonable spirit is let of all its works in the body. As it is seen in them that be amazed, and mad men and frantic, and in others that oft lose use of reason.

The sight is most simple, for it is fiery, and knoweth suddenly things that be full far. The sight is shapen in this manner. In the middle of the eye, that is, the black thereof, is a certain humour most pure and clear. The philosophers call it crystalloid, for it taketh suddenly divers forms and shapes of colours as crystal doth. The sight is a wit of perceiving and knowing of colours, figures, and shapes, and outer properties. Then to make the sight perfect, these things are needful, that is to wit, the cause efficient, the limb of the eye convenient to the thing that shall be seen, the air that bringeth the likeness to the eye, and taking heed, and easy moving. The cause efficient is that virtue that is called animal. The instrument and limb is the humour like crystal in either eye clear and round. It is clear that by the clearness thereof the eye may beshine the spirit, and air; it is round that it be stronger to withstand griefs. The outer thing helping to work, is the air, without which being a means, the sight may not be perfect. It needeth to take heed, for if the soul be occupied about other things than longeth to the sight, the sight is the less perfect. For it deemeth not of the thing that is seen. And easy moving is needful, for if the thing that is seen moveth too swiftly, the sight is cumbered and disparcled with too swift and continual moving: as it is in an oar that seemeth broken in the water, through the swift moving of the water. In three manners the sight is made. One manner by straight lines, upon the which the likeness of the thing that is seen, cometh to the sight. Another manner, upon lines rebounded again: when the likeness of a thing cometh therefrom to a shewer, and is bent, and re-boundeth from the shewer to the sight. The third manner is by lines, the which though they be not bent and rebounded, but stretched between the thing that is seen and the sight: yet they pass not always forthright, but other whiles they blench some whether, aside from the straight way. And that is when divers manners spaces of divers clearness and thickness be put between the sight and the thing that is seen.

Aristotle rehearseth these five mean colours [between white and black] by name, and calleth the first yellow, and the second citrine, and the third red, the fourth purple, and the fifth green.

In the book Meteorics, a little before the end, Aristotle saith that gold, as other metals, hath other matter of subtle brimstone and red, and of quicksilver subtle and white. In the composition thereof is more sadness of brimstone than of air and moisture of quicksilver, and therefore gold is more sad and heavy than silver. In composition of silver is more commonly quicksilver than white brimstone. Then among metals nothing is more sad in substance, or more better compact than gold.

And therefore though it be put in fire, it wasteth not by smoking and vapours, nor lesseth not the weight, and so it is not wasted in fire, but if it be melted with strong heat, then if any filth be therein, it is cleansed thereof. And that maketh the gold more pure and shining. No metal stretcheth more with hammer work than gold, for it stretcheth so, that between the anvil and the hammer without breaking and rending in pieces it stretcheth to gold foil. And among metals there is none fairer in sight than gold, and therefore among painters gold is chief and fairest in sight, and so it embellisheth colour and shape, and colour of other metals. Also among metals is nothing so effectual in virtue as gold. Plato describeth the virtue thereof and saith that it is more temperate and pure than other metals. For it hath virtue to comfort and for to cleanse superfluities gathered in bodies. And therefore it helpeth against leprosy and meselry. The filings of gold taken in meat or in drink or in medicine, preserve and let breeding of leperhood, or namely hideth it and maketh it unknown.

Orpiment is a vein of the earth, or a manner of free stone that cleaveth and breaketh, and it is like to gold in colour: and this is called Arsenic by another name, and is double, red and citron. It hath kind of brimstone, of burning and drying. And if it be laid to brass, it maketh the brass white, and burneth and wasteth all bodies of metal, out take gold.

Though silver be white yet it maketh black lines and strakes in the body that is scored therewith. In composition thereof is quicksilver and white brimstone, and therefore it is not so heavy as gold. There are two manner of silvers, simple and compound. The simple is fleeting, and is called quicksilver; the silver compounded is massy and sad, and is compounded of quicksilver pure and clean, and of white brimstone, not burning, as Aristotle saith.

Quicksilver is a watery substance medlied strongly with subtle earthly things, and may not be dissolved: and that is for great dryness of earth that melteth not on a plain thing. Therefore it cleaveth not to thing that it toucheth, as doth the thing that is watery. The substance thereof is white: and that is for clearness of clear water, and for whiteness of subtle earth that is well digested. Also it hath whiteness of medlying of air with the aforesaid things. Also quicksilver hath the property that it curdeth not by itself kindly without brimstone: but with brimstone, and with substance of lead, it is congealed and fastened together. And therefore it is said, that quicksilver and brimstone is the element, that is to wit matter, of which all melting metal is made. Quicksilver is matter of all metal, and therefore in respect of them it is a simple element. Isidore saith it is fleeting, for it runneth and is specially found in silver forges as it were drops of silver molten. And it is oft found in old dirt of sinks, and in slime of pits. And also it is made of minium done in caverns of iron, and a patent or a shell done thereunder; and the vessel that is anointed therewith, shall be be=clipped with burning coals, and then the quicksilver shall drop. Without this silver nor gold nor latten nor copper may be overgilt. And it is of so great virtue and strength, that though thou do a stone of an hundred pound weight upon quicksilver of the weight of two pounds, the quicksilver anon withstandeth the weight. And if thou doest thereon a scruple of gold, it ravisheth unto itself the lightness thereof. And so it appeareth it is not weight, but nature to which it obeyeth. It is best kept in glass vessels, for it pierceth, boreth, and fretteth other matters.

If an adamant be set by iron, it suffereth not the iron to come to the magnet, but it draweth it by a manner of violence from the magnet, so that though the magnet draweth iron to itself, the adamant draweth it away from the magnet. It is called a precious stone of reconciliation and of love. For if a woman be away from her housebond, or trespasseth against him: by virtue of this stone, she is the sooner reconciled to have grace of her husband.

Crystal is a bright stone and clear, with watery colour. Men trowe that it is of snow or ice made hard in space of many years. This stone set in the sun taketh fire, insomuch if dry tow be put thereto, it setteth the tow on fire. That crystal materi=ally is made of water, Gregory on Ezekiel i. saith: water, saith he, is of itself fleeting, but by strength of cold it is turned and made stedfast crystal. And hereof Aristotle telleth the cause in his Meteorics: there he saith that stony things of substance of ore are water in matter. Ricardus Rufus saith: stone ore is of water: but for it hath more of dryness of earth than things that melt, therefore they were not frozen only with coldness of water, but also by dryness of earth that is mingled therewith, when the watery part of the earth and glassy hath mastery on the water, and the aforesaid cold hath the victory and mastery. And so Saint Gregory his reason is true, that saith, that crystal may be gendered of water.

In old time or the use of iron was known, men eared land with brass, and fought therewith in war and battle. That time gold and silver were forsaken, and gold is now in the most worship, so age that passeth and vadeth changeth times of things. Brass and copper are made in this manner as other metals be, of brimstone and quicksilver, and that happeneth when there is more of brimstone than of quicksilver, and the brimstone is earthy and not pure, with red colour and burning, and quicksilver is mean and not subtle. Of such medlying brass is gendered.

Electrum is a metal and hath that name, for in the sunbeam it shineth more clear than gold or silver. And this metal is more noble than other metals. And hereof are three manners of kinds. The third manner is made of three parts of gold, and of the fourth of silver: and kind electrum is of that kind, for in twinkling and in light it shineth more clear than all other metal, and warneth of venom, for if one dip it therein, it maketh a great chinking noise, and changeth oft into divers colours as the rainbow, and that suddenly.

Heliotrope is a precious stone, and is green, and sprinkled with red drops, and veins of the colour of blood. If it be put in water before the sunbeams, it maketh the water seethe in the vessel that it is in, and resolveth it as it were into mist, and soon after it is resolved into rain-drops. Also it seemeth that this same stone may do wonders, for if it be put in a basin with clear water, it changeth the sunbeams by rebounding of the air, and seemeth to shadow them, and breedeth in the air red and sanguine colour, as though the sun were in eclypse and darkened. An herb of the same name, with certain enchantments, doth beguile the sight of men that look thereon, and maketh a man that beareth it not to be seen.

Though iron cometh of the earth, yet it is most hard and sad, and therefore with beating and smiting it suppresseth and dilateth all other metal, and maketh it stretch on length and on breadth. Iron is gendered of quicksilver thick and not clean, full of earthy holes, and of brimstone, great and boisterous and not pure. In composition of iron is more of the aforesaid brimstone than of quicksilver, and so for mastery of cold and dry and of earthy matter, iron is dry and cold and full well hard, and is compact together in its parts. And for iron hath less of airy and watery moisture than other metals: therefore it is hard to resolve and make it again to be nesh in fire. Use of iron is more needful to men in many things than use of gold: though covetous men love more gold than iron. Without iron the commonalty be not sure against enemies, without dread of iron the common right is not governed; with iron innocent men are defended: and fool-hardiness of wicked men is chastised with dread of iron. And well nigh no handiwork is wrought without iron: no field is eared without iron, neither tilling craft used, nor building builded without iron. And therefore Isidore saith that iron hath its name *ferrum*, for that thereby *farra*, that is corn and seed, is tilled and sown. For, without iron, bread is not won of the earth, nor bread is not departed when it is ready without iron convenably to man's use.

Of lead are two manner of kinds, white and black, and the white is the better, and was first found in the islands of the Atlantic Sea in old time, and is now found in many places. For in France and in Portugal is a manner of black earth found full of gravel and of small stones, and is washed and blown, and so of that matter cometh the substance of lead. Also in gold quarries with matter of gold are small stones found, and are gathered with the gold, and blown by themselves, and turn all to lead, and therefore gold is as heavy as lead. But of black lead is double kind. For black lead cometh alone of a vein, or is gendered of silver in medlied veins, and is blown, and in blowing first cometh tin, and then silver, and then what leaveth is blown and turneth into black lead. Aristotle saith that of brimstone that is boisterous and not swiftly pured, but troublous and thick, and of quicksilver, the substance of lead is gendered, and is gendered in mineral places; so of uncleanness of impure brimstone lead hath a manner of neshness, and smircheth his hand that toucheth it. And with wiping and cleansing, this uncleanness of lead may be taken away for a time, but never for always; a man may wipe off the uncleanness but alway it is lead although it seemeth silver. But strange qualities have mastery therein and beguile men, and make them err therein. Some men take Sal Ammoniac (to cleanse it) as Aristotle saith, and assigneth the cause of this uncleanness and saith, that in boisterous lead is evil quicksilver heavy and fenny. Also that brimstone thereof is evil vapour and stinking. Therefore it freezeth not well at full. Hermes saith that lead in boiling undoeth the hardness of all sad and hard bodies, and also of the stone adamant. Aristotle speaketh of lead in the Meteorics and saith that lead without doubt when it is molten is as quicksilver, but it melteth not without heat, and then all that is molten seemeth red. Wonder it is that though lead be pale or brown, yet by burning or by refudation of vinegar oft it gendereth seemly colour and fair, as tewly, red, and such other; therewith women paint themselves for to seem fair of colour.

The sapphire is a precious stone, and is blue in colour, most like to heaven in fair weather, and clear, and is best among precious stones, and most apt and able to fingers of kings. Its virtue is contrary to venom and quencheth it every deal. And if thou put an addercop in a box, and hold a very sapphire of Ind at the mouth of the box any while, by virtue thereof the addercop is overcome and dieth, as it were suddenly. And this same I have seen proved oft in many and divers places.

Tin in fire departeth metals of divers kind, and it departeth lead and brass from gold and silver, and defendeth other metals in hot fire. And though brass and iron be most hard in kind, yet if they be in strong fire without tin, they burn and waste away. If brazen vessels be tinned, the tin abateth the venom of rust, and amendeth the savour. Also mirrors be tempered with tin, and white colour that is called Ceruse is made of tin, as it is made of lead. Aristotle saith that tin is compounded of good quicksilver and of evil brimstone. And these twain be not well medlied but in small parts compounded, therefore tin hath colour of silver but not the sadness thereof. In the book of Alchemy Hermes saith, that tin breaketh all metals and bodies that it is medlied with, and that for the great dryness of tin. And destroyeth in metal the kind that is obedient to hammer work. And if thou medliest quicksilver therewith, it withstandeth the crassing thereof and maketh it white, but afterward it maketh it black and defileth it. Also there it is said that burnt tin gendereth red colour, as lead doth; and if the fire be strong, the first matter of tin cometh soon again. Also though tin be more nesh than silver, and more hard than lead, yet lead may not be soon soldered to lead nor to brass nor to iron without tin. Neither may these be soldered without grease or tallow.

Brimstone is a vein of the earth and hath much air and fire in its composition. Of brimstone there are four kinds. One is called *vivum*, the which when it is digged, shineth and flourisheth, the which only among all the kinds thereof physicians use. Avicenna means that brimstone is hot and dry in the fourth degree, and is turned into kind of brimstone in part of water, of earth, and of fire, and that brimstone is sometimes great and boisterous and full of drausts, and sometimes pure white, clear and subtle, and sometimes mean between both. And by this diverse disposition, divers metals are gendered of brimstone and of quicksilver.

Glass, as Avicen saith, is among stones as a fool among men, for it taketh all manner of colour and painting. Glass was first found beside Ptolomeida in the cliff beside the river that is called Vellus, that springeth out of the foot of Mount Carmel, at which shipmen arrived. For upon the gravel of that river shipmen made fire of clods medlied with bright gravel, and thereof ran streams of new liquor, that was the beginning of glass. It is so pliant that it taketh anon divers and contrary shapes by blast of the glazier, and is sometimes beaten, and sometimes graven as silver. And no matter is more apt to make mirrors than is glass, or to receive painting; and if it be broken it may not be amended without melting again. But long time past, there was one that made glass pliant, which might be amended and wrought with an hammer, and brought a vial made of such glass tofore Tiberius the Emperor, and threw it down on the ground, and it was not broken but bent and folded. And he made it right and amended it with an hammer. Then the emperor commanded to smite off his head anon, lest that his craft were known. For then gold should be no better than fen, and all other metal should be of little worth, for certain if glass vessels were not brittle, they should be accounted of more value than vessels of gold.

All the planets move by double moving; by their own kind moving out of the west into the east, against the moving of the firmament; and by other moving out of the east into the west, and that by ravishing of the firmament. By violence of the firmament they are ravished every day out of the east into the west. And by their kindly moving, by the which they labour to move against the firmament, some of them fulfil their course in shorter time, and some in longer time. And that is for their courses are some more and some less. For Saturn abideth in every sign xxx months, and full endeth its course in xxx years. Jupiter dwelleth in every sign one year, and full endeth its course in xii years. Mars abideth in every sign xlv days, and full endeth its course in two years. The sun abideth in every sign xxx days and ten hours and a half, and full endeth its course in ccclxv days and vi hours. Mercury abideth in every sign xxviii days and vi hours, and full endeth its course in cccxxxviii days. Venus abideth in every sign 29 days, and full endeth its course in 348 days. The moon abideth in every sign two days and a half, and six hours and one bisse less, and full endeth its course from point to point in 27 days and 8 hours. And by entering and out passing of these 7 stars into the 12 signs and out thereof everything that is bred and corrupt in this nether world is varied and disposed, and therefore in the philosopher's book Mesalath it is read in this manner: "The Highest made the world to the likeness of a sphere, and made the highest circle above it moveable in the earth, pight and stedfast in the middle thereof; not withdrawing toward the left side, nor toward the right side, and set the other elements moveable, and made them move by the moving of 7 planets, and all other stars help the planets in their working and kind." Every creature upon Earth hath a manner inclination by the moving of the planets, and destruction cometh by moving and working of planets. The working of them varieth and is diverse by diversity of climates and countries. For they work one manner of thing about the land of blue men, and another about the land and country of Slavens.... In the signs the planets move and abate with double moving, and move by accidental ravishing of the firmament out of the East into the West; and by kindly moving, the which is double, the first and the second. The first moving is the round moving that a planet maketh in its own circle, and passeth never the marks and bounds of the circle. The second moving is that he maketh under the Zodiac, and passeth alway like great space in a like space of time. And the first moving of a planet is made in its own circle that is called Eccentric, and it is called so for the earth is not the middle thereof, as it is the middle of the circle that is called Zodiac. Epicycle is a little circle that a planet describeth, and goeth about therein by the moving of its body, and the body of the planet goeth about the roundness thereof. And therefore it sheweth, that the sun and other planets move in their own circles; and first alike swift, though they move diversely in divers circles. Also in these circles the manner moving of planets is full wisely found of astronomers, that are called Direct, Stationary, and Retrograde Motion. Forthright moving is in the over part of the circle that is called Epicycle, backward is in the nether part, and stinting and abiding or hoving is in the middle.

## II
## MEDIAEVAL MANNERS

The sixth book of our author deals with the conditions of man, passing in review youth and age, male and female, serf and lord. Our extracts from it fall into three groups. The first deals in great measure with the relations of family life. We have an account of the boy and the girl (as they appeared to a friar "of orders grey"), the infant and its nurse. However we may suspect Bartholomew of wishing to provide a text in his account of the bad boy, it is consoling to find that the "enfant terrible" had his counterpart in the thirteenth century, as well as the maiden known to us all, who is "demure and soft of speech, but well ware of what she says."

The second group presents mediaeval society to us under the influence of chivalry. Suitably enough, we have beside each other most lifelike pictures of the base and superstructure of the system. This, the man-- free, generous; that, the serf--vile, ungrateful, kept in order by fear alone, but the necessary counterpart of the splendid figure of his master. One of our writers today has regretted the absence of a chapter in praise of the good man to set beside Solomon's picture of the virtuous woman. Bartholomew has certainly endeavoured in the two chapters quoted here, "Of a Man," and "Of a Good Lord," to picture the ideal good man of chivalrous times. It may, however, be permitted those of us who look at the system from underneath, to sympathise with our fellows who struggled to free themselves from bondage under Tyler and John Ball at least as much as with their splendid oppressors, and to recognise that the feudal system, however necessary in the thirteenth century, lost its value when its lords had ceased to be such good lords as our author describes.

The third group would naturally consist of passages illustrating the daily life of our ancestors, but the editor has found some difficulty in getting together passages enough for the purpose without trenching on the confines of other chapters. He has accordingly left them scattered over the book, persuaded that the reader will feel their import better when they are seen in their context. Such a book as this is not open to the objections urged against pictures of mediaeval life drawn from romances, that the situations are invented and the manners suited to the situation. Here all is true, and written with no other aim than that of utilising knowledge common to all. Everywhere through these extracts little statements--a few words in most cases--crop up giving us information of this kind; but it would be impossible to do more than allude to them. Leaving our reader to notice them as they are met with, the description of a mediaeval dinner concludes the chapter. The chapter describing a supper which follows it in the original is too long for quotation, and is vitiated by a desire to draw analogies. But one feature is noteworthy: Among the properties of a good supper, "the ninth is plenty of light of candles, and of prickets, and of torches. For it is shame to sup in darkness, and perillous also for flies and other filth. Therefore candles and prickets are set on candlesticks and chandeliers, lanterns and lamps are necessary to burn." This little touch gives us the reverse of the picture, and reminds us of the Knight of the Tower's caution to his daughters about their behaviour at a feast.

SUCH children be nesh of flesh, lithe and pliant of body, able and light to moving, witty to learn. And lead their lives without thought and care. And set their courages only of mirth and liking, and dread no perils more than beating with a rod: and they love an apple more than gold. When they be praised, or shamed, or blamed, they set little thereby. Through stirring and moving of the heat of the flesh and of humours, they be lightly and soon wroth, and soon pleased, and lightly they forgive. And for tenderness of body they be soon hurt and grieved, and may not well endure hard travail. Since all children be tatched with evil manners, and think only on things that be, and reck not of things that shall be, they love plays, game, and vanity, and forsake winning and profit. And things most worthy they repute least worthy, and least worthy most worthy. They desire things that be to them contrary and grievous, and set more of the image of a child, than of the image of a man, and make more sorrow and woe, and weep more for the loss of an apple, than for the loss of their heritage. And the goodness that is done for them, they let it pass out of mind. They desire all things that they see, and pray and ask with voice and with hand. They love talking and counsel of such children as they be, and void company of old men. They keep no counsel, but they tell all that they hear or see. Suddenly they laugh, and suddenly they weep. Always they cry, jangle, and jape; that unneth they be still while they sleep. When they be washed of filth, anon they defile themselves again. When their mother washeth and combeth them, they kick and sprawl, and put with feet and with hands, and withstand with all their might. They desire to drink always, unneth they are out of bed, when they cry for meat anon.

Men behove to take heed of maidens: for they be tender of complexion; small, pliant and fair of disposition of body: shamefast, fearful, and merry. Touching outward disposition they be well nurtured, demure and soft of speech, and well ware of what they say: and delicate in their apparel. And for a woman is more meeker than a man, she weepeth sooner. And is more envious, and more laughing, and loving, and the malice of the soul is more in a woman than in a man. And she is of feeble kind, and she maketh more lesings, and is more shamefast, and more slow in working and in moving than is a man.

A nurse hath that name of nourishing, for she is ordained to nourish and to feed the child, and therefore like as the mother, the nurse is glad if the child be glad, and heavy, if the child be sorry, and taketh the child up if it fall, and giveth it suck: if it weep she kisseth and lulleth it still, and gathereth the limbs, and bindeth them together, and doth cleanse and wash it when it is defiled. And for it cannot speak, the nurse lispeth and soundeth the same words to teach more easily the child that cannot speak. And she useth medicines to bring the child to convenable estate if it be sick, and lifteth it up now on her shoulders, now on her hands, now on her knees and lap, and lifteth it up if it cry or weep. And she cheweth meat in her mouth, and maketh it ready to the toothless child, that it may the easilier swallow that meat, and so she feedeth the child when it is an hungered, and pleaseth the child with whispering and songs when it shall sleep, and swatheth it in sweet clothes, and righteth and stretcheth out its other. A man hath so great love to his wife that for her sake he adventureth himself to all perils; and setteth her love afore his mother's love; for he dwelleth with his wife, and forsaketh father and mother. Afore wedding, the spouse thinketh to win love of her that he wooeth with gifts, and certifieth of his will with letters and messengers, and with divers presents, and giveth many gifts, and much good and cattle, and promiseth much more. And to please her he putteth him to divers plays and games among gatherings of men, and useth oft deeds of arms, of might, and of mastery. And maketh him gay and seemly in divers clothing and array. And all that he is prayed to give and to do for her love, he giveth and doth anon with all his might. And denieth no petition that is made in her name and for her love. He speaketh to her pleasantly, and beholdeth her cheer in the face with pleasing and glad cheer, and with a sharp eye, and at last assenteth to her, and telleth openly his will in presence of her friends, and spouseth her with a ring, and giveth her gifts in token of contract of wedding, and maketh her charters, and deeds of grants and of gifts. He maketh revels and feasts and spousals, and giveth many good gifts to friends and guests, and comforteth and gladdeth his guests with songs and pipes and other minstrelsy of music. And afterward, when all this is done, he bringeth her to the privities of his chamber, and maketh her fellow at bed and at board. And then he maketh her lady of his money, and of his house, and meinie. And then he is no less diligent and careful for her than he is for himself: and specially lovingly he adviseth her if she do amiss, and taketh good heed to keep her well, and taketh heed of her bearing and going, of her speaking and looking, of her passing and ayencoming, out and home. No man hath more wealth, than he that hath a good woman to his wife, and no man hath more woe, than he that hath an evil wife, crying and jangling, chiding and scolding, drunken, lecherous, and unsteadfast, and contrary to him, costly, stout and gay, envious, noyful, leaping over lands, much suspicious, and wrathful. In a good spouse and wife behoveth these conditions, that she be busy and devout in God's service, meek and serviceable to her husband, and fair- speaking and goodly to her meinie, merciful and good to wretches that be needy, easy and peaceable to her neighbours, ready, wary, and wise in things that should be avoided, mightiful and patient in suffering, busy and diligent in her doing, mannerly in clothing, sober in moving, wary in speaking, chaste in looking, honest in bearing, sad in going, shamefast among the people, merry and glad with her husband, and chaste in privity. Such a wife is worthy to be praised, that entendeth more to please her husband with such womanly dues, than with her braided hairs, and desireth more to please him with virtues than with fair and gay clothes, and useth the goodness of matrimony more because of children than of fleshly liking, and hath more liking to have children of grace than of kind.

A man loveth his child and feedeth and nourisheth it, and setteth it at his own board when it is weaned. And teacheth him in his youth with speech and words, and chasteneth him with beating, and setteth him and putteth him to learn under ward and keeping of wardens and tutors. And the father sheweth him no glad cheer, lest he wax proud, and he loveth most the son that is like to him, and looketh oft on him. And giveth to his children clothing, meat and drink as their age requireth, and purchaseth lands and heritage for his children, and ceaseth not to make it more and more. And entaileth his purchase, and leaveth it to his heirs.... The child cometh of the substance of father and mother, and taketh of them feeding and nourishing, and profiteth not, neither liveth, without help of them. The more the father loveth his child, the more busily he teacheth and chastiseth him and holdeth him the more strait under chastising and lore; and when the child is most loved of the father it seemeth that he loveth him not; for he beateth and grieveth him oft lest he draw to evil manners and tatches, and the more the child is like to the father, the better the father loveth him. The father is ashamed if he hear any foul thing told by his children. The father's heart is sore grieved, if his children rebel against him. In feeding and nourishing of their children stands the most business and charge of the parents.

Some servants be bond and born in bondage, and such have many pains by law. For they may not sell nor give away their own good and cattle, nother make contracts, nother take office of dignity, nother bear witness without leave of their lords. Wherefore though they be not in childhood, they be oft punished with pains of childhood. Other servants there be, the which being taken with strangers and aliens and with enemies be bought and sold, and held low under the yoke of thraldom. The third manner of servants be bound freely by their own good will, and serve for reward and for hire. And these commonly be called Famuli.

The name lord is a name of sovereignty, of power, and of might. For without a lord might not the common profit stand secure, neither the company of men might be peaceable and quiet. For if power and might of rightful lords were withholden and taken away, then were malice free, and goodness and innocence never secure, as saith Isidore. A rightful lord, by way of rightful law, heareth and determineth causes, pleas, and strifes, that be between his subjects, and ordaineth that every man have his own, and draweth his sword against malice, and putteth forth his shield of righteousness, to defend innocents against evil doers, and delivereth small children and such as be fatherless, and motherless, and widows, of them that overset them. And he pursueth robbers and rievers, thieves, and other evil doers. And useth his power not after his own will, but he ordaineth and disposeth it as the law asketh.... By reason of one good king and one good lord, all a country is worshipped, and dreaded, and enhanced also. Also this name lord is a name of peace and surety. For a good lord ceaseth war, battle, and fighting; and accordeth them that be in strife. And so under a good, a strong, and a peaceable lord, men of the country be secure and safe. For there dare no man assail his lordship, ne in no manner break his peace.

Meat and drink be ordained and convenient to dinners and to feasts, for at feasts first meat is prepared and arrayed, guests be called together, forms and stools be set in the hall, and tables, cloths, and towels be ordained, disposed, and made ready. Guests be set with the lord in the chief place of the board, and they sit not down at the board before the guests wash their hands. Children be set in their place, and servants at a table by themselves. First knives, spoons, and salts be set on the board, and then bread and drink, and many divers messes; household servants busily help each other to do everything diligently, and talk merrily together. The guests be gladded with lutes and harps. Now wine and now messes of meat be brought forth and departed. At the last cometh fruit and spices, and when they have eaten, board, cloths, and relief are borne away, and guests wash and wipe their hands again. Then grace is said, and guests thank the lord. Then for gladness and comfort drink is brought yet again. When all this is done at meat, men take their leave, and some go to bed and sleep, and some go home to their own lodgings.

# III
# MEDIAEVAL MEDICINE

The seventh book of the "De Proprietatibus" treats of the human body and its ailments. At first glance it might seem that such a subject would be repulsive, either in matter or handling, to the general reader of today, but it will, we think, be found that there are many points of interest in it for us, some of which we proceed to indicate. Mankind has always felt a deep interest in certain diseases, to which we are even now subject, and so parts of the chapters on leprosy and hydrophobia have been reproduced. The accounts given of frenzy and madness interest us both as a picture of the change in manners, as an example of the methods of cure proposed, and as throwing light on many passages. Thus Chaucer, speaking of Arcite, describes his passion as compounded of melancholy which deprives him of reason, overflowing into the foremost cell of his brain, the cell fantastic, and causing him to act as if mad.

"Nought oonly lyke the loveres maladye
Of Hereos, but rather lyk manye,
Engendered of humour malencolyk
Byforen in his selle fantastyk."
K. T., 515, etc.

Physicians recommend music as a cure in mental troubles, but that it is no new discovery is attested by Shakespeare and our author. Compare what Bartholomew says of the voice, with Richard's speech:

"This music mads me, let it sound no more,

For though it have holp madmen to their wits,
In me it seems it will make wise men mad."

The origin of the brutality towards madmen warred against by Charles Reade, and described in "Romeo and Juliet"--

"Not mad, but bound more than a madman is,

Shut up in prison, kept without my food,
Whipp'd and tormented"--
is seen in our extracts, which recall, too, in their insistence on bleeding the "head vein," Juvenal's remark on his friend about to marry: "O medici, mediam pertundite venam."

Some space has already been devoted (p. 28) to the physiology of the human body, but this chapter would not be complete if we did not devote some space to the explanations given of the working of the heart, veins, and arteries, at a time when the circulation of the blood was unknown. It may not be amiss to remind the reader that arteries carry blood from the heart, to which it is returned by the veins, after passing through a fine network of tubes called the capillaries.

Turning to what may be called the popular physiology of the time, we may note the change, since mediaeval times, in the allocation of properties to the organs of the body. In our days, the heart and brain set aside, we find no organ mentioned in connection with the various faculties of the body, while up to Shakespeare's time each organ had its passion. Some of these emotions have much changed their seats. True love, which now reigns over the heart, then took its rise in the liver. The friar in "Much Ado about Nothing" says of Claudio, "If ever love had interest in his liver"; and the Duke in "Twelfth Night," speaking of women's love, says:

"Alas, their love may be call'd appetite,
No motion of the liver, but the palate."

The heart, on the other hand, was considered as the seat of wisdom.
The spleen is now almost a synonym for bitterness of spirit, but it used to be

regarded as the source of laughter. Isabella in "Measure for Measure," after the well-known quotation about man dressed in a little brief authority who plays such apish tricks as make the angels weep, says they would laugh instead if they had spleens:

"Who, with our spleens,
Would all themselves laugh mortal."

The brain in mediaeval times was regarded only as the home of the "wits of feeling"--the senses.

Some other points of interest in mediaeval medicine are the strange remedies prescribed, and the way in which they were hit upon. The Editor has not made many selections to illustrate this, nor has he sought out the most strange. And lastly, in this, as in most of the other chapters, much may be learnt of the customs of the time from the indications of the text.

These be the signs of frenzy, woodness and continual waking, moving and casting about the eyes, raging, stretching, and casting out of hands, moving and wagging of the head, grinding and gnashing together of the teeth; always they will arise out of their bed, now they sing, now they weep, and they bite gladly and rend their keeper and their leech: seldom be they still, but cry much. And these be most perilously sick, and yet they wot not then that they be sick. Then they must be soon holpen lest they perish, and that both in diet and in medicine. The diet shall be full scarce, as crumbs of bread, which must many times be wet in water. The medicine is, that in the beginning the patient's head be shaven, and washed in lukewarm vinegar, and that he be well kept or bound in a dark place. Diverse shapes of faces and semblance of painting shall not be shewed tofore him, lest he be tarred with woodness. All that be about him shall be commanded to be still and in silence; men shall not answer to his nice words. In the beginning of medicine he shall be let blood in a vein of the forehead, and bled as much as will fill an egg-shell. Afore all things (if virtue and age suffereth) he shall bleed in the head vein. Over all things, with ointments and balming men shall labour to bring him asleep. The head that is shaven shall be plastered with lungs of a swine, or of a wether, or of a sheep; the temples and forehead shall be anointed with the juice of lettuce, or of poppy. If after these medicines are laid thus to, the woodness dureth three days without sleep, there is no hope of recovery.

Madness is infection of the foremost cell of the head, with privation of imagination, like as melancholy is the infection of the middle cell of the head, with privation of reason.

Madness cometh sometime of passions of the soul, as of business and of great thoughts, of sorrow and of too great study, and of dread: sometime of the biting of a wood hound, or some other venomous beast: sometime of melancholy meats, and sometime of drink of strong wine. And as the causes be diverse, the tokens and signs be diverse. For some cry and leap and hurt and wound themselves and other men, and darken and hide themselves in privy and secret places. The medicine of them is, that they be bound, that they hurt not themselves and other men. And namely, such shall be refreshed, and comforted, and withdrawn from cause and matter of dread and busy thoughts. And they must be gladded with instruments of music, and somedeal be occupied.

Our Lord set a token in Cain, that was quaking of head, as Strabus saith in the gloss: "Every man (saith Strabus) that findeth me, by quaking of head and moving of wood heart, shall know that I am guilty to die."

Among all the passions and evils of the wits of feeling, blindness is most wretched. For without any bond, blindness is a prison to the blind. And blindness beguileth the virtue imaginative in knowing; for in deeming of white the blind deem it is black, and ayenward. It letteth the virtue of avisement in deeming. For he deemeth and aviseth, and casteth to go eastward, and is beguiled in his doom, and goeth westward. And blindness over-turneth the virtue of affection and desire. For if men proffer the blind a silver penny and a copper to choose the better, he desireth to choose the silver penny, but he chooseth the copper.

The blind man's wretchedness is so much, that it maketh him not only subject to a child, or to a servant, for ruling and leading, but also to an hound. And the blind is oft brought to so great need, that to pass and scape the peril of a bridge or of a ford, he is compelled to trust in a hound more than to himself. Also oft in perils where all men doubt and dread, the blind man, for he seeth no peril, is secure. And in like wise there as is no peril, the blind dreadeth most. He spurneth oft in plain way, and stumbleth oft; there he should heave up his foot, he boweth it downward. And in like wise there as he should set his foot to the ground, he heaveth it upward. He putteth forth the hand all about groping and grasping, he seeketh all about his way with his hand and with his staff. Seldom he doth aught securely, well nigh always he doubteth and dreadeth. Also the blind man when he lieth or sitteth thereout, he weeneth that he is under covert; and ofttimes he thinketh himself hid when everybody seeth him.

Also sometimes the blind beateth and smiteth and grieveth the child that leadeth him, and shall soon repent the beating by doing of the child. For the child hath mind of the beating, and forsaketh him, and leaveth him alone in the middle of a bridge, or in some other peril, and teacheth him not the way to void the peril. Therefore the blind is wretched, for in house he dare nothing trustly do, and in the way he dreadeth lest his fellow will forsake him.

Universally this evil [leprosy] hath much tokens and signs. In them the flesh is notably corrupt, the shape is changed, the eyen become round, the eyelids are revelled, the sight sparkleth, the nostrils are straited and revelled and shrunk. The voice is hoarse, swelling groweth in the body, and many small botches and whelks hard and round, in the legs and in the utter parts; feeling is somedeal taken away. The nails are boystous and bunchy, the fingers shrink and crook, the breath is corrupt, and oft whole men are infected with the stench thereof. The flesh and skin is fatty, insomuch that they may throw water thereon, and it is not the more wet, but the water slides off, as it were off a wet hide. Also in the body be diverse specks, now red, now black, now wan, now pale. The tokens of leprosy be most seen in the utter parts, as in the feet, legs, and face; and namely in wasting and minishing of the brawns of the body.

To heal or to hide leprosy, best is a red adder with a white womb, if the venom be away, and the tail and the head smitten off, and the body sod with leeks, if it be oft taken and eaten. And this medicine helpeth in many evils; as appeareth by the blind man, to whom his wife gave an adder with garlick instead of an eel, that it might slay him, and he ate it, and after that by much sweat, he recovered his sight again.

The biting of a wood hound is deadly and venomous. And such venom is perilous. For it is long hidden and unknown, and increaseth and multiplieth itself, and is sometimes unknown to the year's end, and then the same day and hour of the biting, it cometh to the head, and breedeth frenzy. They that are bitten of a wood hound have in their sleep dreadful sights, and are fearful, astonied, and wroth without cause. And they dread to be seen of other men, and bark as hounds, and they dread water most of all things, and are afeared thereof full sore, and squeamous also. Against the biting of a wood hound wise men and ready used to make the wounds bleed with fire or with iron, that the venom may come out with blood, that cometh out of the wound.

Then consider thou shortly hereof, that a physician visiteth oft the houses and countries of sick men. And seeketh and searcheth the causes and circumstances of the sicknesses, and arrayeth and bringeth with him divers and contrary medicines. And he refuseth not to grope and handle, and to wipe and cleanse wounds of sick men. And he behooteth to all men hope and trust of recovering of health; and saith that he will softly burn that which shall be burnt, and cut that which shall be cut. And lest the whole part should corrupt, he spareth not to burn and to cut off the part that is rotted, and if a part in the right side acheth, he spareth not to smite in the left side. A good leech leaveth not cutting or burning for weeping of the patient. And he hideth and covereth the bitterness of the medicine with some manner of sweetness. He drinketh and tasteth of the medicine, though it be bitter: that it be not against the sick man's heart, and refraineth the sick man of meat and drink; and letteth him have his own will, of the whose health is neither hope nor trust of recovering.

The veins have that name for that they be the ways, conduits, and streams of the fleeting of the blood, and sheddeth it into all the body. And Constantine saith, that the veins spring out of the liver, as the arteries and wosen do out of the heart, and the sinews out of the brain. And veins are needful as vessels of the blood to bear and to bring blood from the liver, to feed and nourish the members of the body. Also needly, the veins are more tender and nesh in kind than sinews. Therefore that they be nigh to the liver may somewhat change the blood that cometh to them. And all the veins are made of one curtel, and not of two, as the arteries and wosen. For the arteries receive spirits, and they keep and save them. And the veins coming out of the liver, suck thereof, as it were of their own mother, feeding of blood, and dealeth and departeth that feeding to every member as it needeth. And so the veins spread into all the parts of the body, and by a wonder wit of kind, they do service each to other.

Also among other veins open and privy, there is a vein, and it is called Artery, which is needful in kind to bear and bring kindly heat from the heart to all the other members. And these arteries are made and composed of two small clothings or skins, called curtels, and they be like in shape, and divers in substance. The inner have wrinkles and folding overthwart, and their substance is hard, and more boystous than the utter be. And without they have wrinkles and folding in length: of whom the substance is hard for needfulness of moving, opening, and closing. For by opening, itself doth receive from the heart and that by the wrinklings and folding in length; by closing, itself doth put out superfluous fumosity, which is done by wrinkling and folding the curtels overthwart and in breadth, in the which the spirit is drawn from the heart. Wherefore they be harder without than all the other veins, and that is needful lest they break lightly and soon. Also these veins spring out of the left hollowness of the heart. And twain of that side are called Pulsative, of which one that is the innermost hath a nesh skin, and this vein is needful to bring great quantity of blood and spirits to the lungs, and to receive in air, and to medley it with blood, to temper the ferventness of the blood. This vein entereth into the lungs and is departed there in many manner wises.

The other artery is more than the first, and Aristotle calleth it Horren; this artery cometh up from the heart, and is departed in twain, and the one part cometh upward, and carrieth blood, that is purified and spirit of life to the brain; that so the spirit of feeling may be bred, nourished, kept, and saved. The other part goeth downward, and is departed in many manner wise toward the right side and toward the left.

Then mark well, that a vein is the bearer and carrier of blood, keeper and warden of the life of beasts. And containeth in itself the four bloody humours clean and pure, which are ordained for feeding of all the parts of the body. Moreover, a vein is hollow to receive blood the more easily, and as it needeth in kind, that one vein bring and give blood to another vein. Also a vein is messager of health and of sickness. For by the pulse of the arteries and disposition of the veins, physicians deem of the feebleness and strength of the heart. Also if a vein be corrupt, and containeth corrupt blood, it corrupteth and infecteth all the body, as it fareth in lepers, whose blood is most corrupt in the veins, of the which the members are fed by sucking of blood, and seeketh thereby corruption and sickness incurable. Also the vein of the arm is oft grieved, constrained and wranged, opened and slit, and wounded, to relieve the sickness of all the body by hurting of that vein.

The spittle of a man fasting hath a manner strength of privy infection. For it grieveth and hurteth the blood of a beast, if it come into a bleeding wound, and is medlied with the blood. And that, peradventure, is, as saith Avicenna, by reason of rawness. For raw humour medlied with blood that hath perfect digestion, is contrary thereto in its quality, and disturbeth the temperance thereof, as authors say. And therefore it is that holy men tell that the spittle of a fasting man slayeth serpents and adders, and is venom to venomous beasts, as saith Basil.

A discording voice and an inordinate troubleth the accord of many voices. But according voices sweet and ordinate, gladden and move to love, and show out the passions of the soul, and witness the strength and virtue of the spiritual members, and show pureness and good disposition of them, and relieve travail, and put off disease and sorrow. And make to be known the male and the female, and get and win praising, and change the affection of the hearers; as it is said in fables of one Orpheus, that pleased trees, woods, hills, and stones, with sweet melody of his voice. Also a fair voice is according and friendly to kind. And pleaseth not only men but also brute beasts, as it fareth in oxen that are excited to travail more by sweet song of the herd, than by strokes and pricks.

Also by sweet songs of harmony and accord or music, sick men and frantic come oft to their wit again and health of body. Some men tell that Orpheus said, "Emperors pray me to feasts, to have liking of me; but I have liking of them which would bend their hearts from wrath to mildness, from sorrow to gladness, from covetousness to largeness, from dread to boldness." This is the ordinance of music, that is known above the sweetness of the soul.

Now it is known by these foresaid things, how profitable is a merry voice and sweet. And contrariwise is of an unordinate voice and horrible, that gladdeth not, nother comforteth; but is noyful and discomforteth and grieveth the ears and the wit. Therefore Constantine saith that a philosopher was questioned, why an horrible man is more heavy than any burden or wit. And men say that he answered in this manner. An horrible man is burden to the soul and wit.

The lungs be the bellows of the heart. It beateth in opening of itself that it may take in breath, and thrusting together may put it out, and so it is in continual moving, in drawing in and out of breath. The lungs be the proper instrument of the heart, for it keleth the heart, and by subtlety of its substance, changeth the air that is drawn in, and maketh it more subtle. The lungs shapeth the voice, and ceaseth never of moving. For it closeth itself and spreadeth, and keepeth the air to help the heat in its dens and holes. And therefore a beast may not live under the water without stifling, but as long as he may hold in the air that is gathered within. The lungs by continual moving putteth off air that is gathered within, cleanseth and purgeth it, and ministereth continual and convenable feeding to the vital spirit. And departeth the heart from the instruments of feeling, and breedeth foamy humours, and beclippeth aside half the substance of the heart. And when the lungs be grieved by any occasion, it speedeth to death- ward.

The liver hath name, for fire hath place therein, that passeth up anon to the brain, and cometh thence to the eyen, and to the other wits and limbs. And the liver by its heat, draweth woose and juice and turneth it into blood, and serveth the body and members therewith, to the use of feeding. In the liver is the place of voluptuousness and liking of the flesh. The ends of the liver hight fibra, for they are straight and passing as tongs, and beclip the stomach, and give heat to digestion of meat: and they hight fibra, because the necromancers brought them to the altars of their god Phoebus and offered them there, and then they had answers.

The liver is the chief fundament of kindly virtue, and greatest helper of the first digestion in the stomach, and the liver maketh perfectly the second digestion in the stomach, in the hollowness of its own substance, and departeth clean and pured, from unclean and unpured, and sendeth feeding to all the members, and exciteth love or bodily lust, and receiveth divers passions. Then the liver is a noble and precious member, by whose alteration the body is altered, and the liver sendeth feeding and virtues of feeding to the other members, to the nether without mean, and to the other, by mean of the heart.

Some men ween, that the milt is cause of laughing. For by the spleen we are moved to laugh, by the gall we are wroth, by the heart we are wise, by the brain we feel, by the liver we love.

# IV
# MEDIAEVAL GEOGRAPHY

The fourteenth and fifteenth books of the "De Proprietatibus" are treatises on the geography of the time. Very few words of the editor's are needed to introduce them to modern readers. They may be divided into two classes: one, interesting because of the legends they preserve for us, the other, as reflecting the social life of the time. The first class is represented here by the accounts of the Amazons, of India, of Ireland, and of Finland. Here we have the outlines of the stories--

> "Of antres vast, and deserts idle,
> Rough quarries, rocks, and hills whose heads touch heaven,
> And of the Cannibals that each other eat,
> The Anthropophagi, and men whose heads
> Do grow beneath their shoulders"--

told by Othello to Desdemona.

In the other we class such accounts as those of France and of Paris, of the Frisians, Flanders, Scotland, and Iceland. Such countries as these were well known in the thirteenth century, and the feelings of our author about them can be gathered easily enough. The tone of the chapters about England and Scotland would be enough alone to prove that Bartholomew was an Englishman, it there were no other reason to think it.

THERE is a lake that hight lake Asphaltus, and is also called the Dead Sea for its greatness and deepness: for it breedeth, ne receiveth, no thing that hath life. Therefore it hath nother fish ne fowls, but whensoever thou wouldst have drowned therein anything that hath life with any craft or gin, then anon it plungeth and cometh again up; though it be strongly thrust downward, it is anon smitten upward. And it moveth not with the wind, for glue withstandeth wind and storms, by which glue all [the] water is stint. And therein may no ship row nor sail, for all thing that hath no life sinketh down to the ground; nor he sustaineth no kind, but it be glued. And a lantern without its light sinketh therein, as it telleth, and a lantern with light floateth above.

As the Master of Histories saith, this lake casteth up black clots of glue. In the brim thereof trees grow, the apples whereof are green till they are ripe: and if ye cut them when they are ripe, ye shall find ashes within them. And so it is said in the gloss; and there grow most fair apples, that make men that see them have liking to eat of them, and if one take them, they fade and fall in ashes and smoke, as though they were burning.

Olympus is a mount of Macedon, and is full high, so that it is said, that the clouds are thereunder, as Virgil saith. This mount departeth Macedonia and Thracia, and is so high, that it passeth all storms and other passions of the air. And therefore philosophers went up to see the course and places of stars, and they might not live there, but if they had sponges with water to make the air more thick by throwing and sprinkling of water: as the Master of Histories saith.

Amazonia, women's land, is a country part in Asia and part in Europe, and is nigh to Albania, and hath that name of Amazonia, of women that were the wives of the men that were called Goths, the which men went out of the nether Scythia, and were cruelly slain, and then their wives took their husbands' armour and weapons, and resed on the enemies with manly hearts, and took wreck of the death of their husbands. For with dint of sword they slew all the young males, and old men, and children, and saved the females, and departed prey, and purposed to live ever after without company of males. And by ensample of their husbands that had alway two kings over them, these women ordained them two queens, that one hight Marsepia, and that other Lampeta, that one should travail with a host, and fight against enemies, and that other should in the mean time, govern and rule the communities. And they were made so fierce warriors in short time, that they had a great part of Asia under their lordship nigh a hundred years: among them they suffered no male to live nor abide, in no manner of wise. But of nations that were nigh to them, they chose husbands because of children, and went to them in times that were ordained, and when the time was done, then they would compel their lovers to go from them, and get other places to abide in, and would slay their sons, or send them to their fathers in certain times. And they saved their daughters, and taught them to shoot and to hunt. And for the shooting of arrows should not be let with great breasts, in the 7th year (as it is said), they burnt off their breasts, and therefore they were called Amazons. And as it is said, Hercules adaunted first the fierceness of them, and then Achilles. But that was more by friendship than by strength, as it is contained in deeds and doings of the Greeks, and the Amazons were destroyed and brought to death by great Alexander. But the story of Alexander saith not so. But it is said that Alexander demanded tribute of the Queen of the Amazons, and she wrote to him again by messengers in this manner.

"Of thy wit I wonder, that thou purposest to fight with women, for if fortune be on our side, and if it hap that thou be overcome, then art thou shamed for evermore, when thou art overcome of women, and if our gods be wroth with us, and thou overcomest us, it shall turn thee to little worship, that thou have the mastery of women."

The noble king wondered on her answer, and said, that it is not seemly to overcome women with sword and with woodness, but rather with fairness and with love: and therefore he granted them freedom and made them subject to his empire, not with violence but with friendship and with love.

England is the most island of Ocean, and is beclipped all about by the sea, and departed from the roundness of the world, and hight sometimes Albion: and had that name of white rocks, which were seen on the sea cliffs. And by continuance of time, lords and noble men of Troy, after that Troy was destroyed, went from thence, and were accompanied with a great navy, and fortuned to the cliffs of the foresaid island, and that by revelation of their feigned goddess Pallas, as it is said, and the Trojans fought with giants long time that dwelled therein, and overcame the giants, both with craft and with strength, and conquered the island, and called the land Britain, by the name of Brute that was prince of that host: and so the island hight Britain, as it were an island conquered of Brute that time, with arms and with might. Of this Brute's offspring came most mighty kings. And who that hath liking to know their deeds, let him read the story of Brute.

And long time after, the Saxons won the island with many and divers hard battles and strong, and their offspring had possession after them of the island, and the Britons were slain or exiled, and the Saxons departed the island among them, and gave every province a name, by the property of its own name and nation, and therefore they cleped the island Anglia, by the name of Engelia the queen, the worthiest duke of Saxony's daughter, that had the island in possession after many battles. Isidore saith, that this land hight Anglia, and hath that name of Angulus, a corner, as it were land set in the end, or a corner of the world. But saint Gregory, seeing English children to sell at Rome, when they were not christened, and hearing that they were called English: according with the name of the country, he answered and said: Truly they be English, for they shine in face right as angels: it is need to send them message, with word of salvation. For as Beda saith, the noble kind of the land shone in their faces. Isidore saith, Britain, that now hight Anglia, is an island set afore France and Spain, and containeth about 48 times 75 miles. Also therein be many rivers and great and hot wells. There is great plenty of metals, there be enough of the stones Agates, and of pearls, the ground is special good, most apt to bear corn and other good fruit. There be, namely, many sheep with good wool, there be many harts and other wild beasts; there be few wolves or none, therefore there be many sheep, and may be securely left without ward, in pasture and in fields, as Beda saith.

England is a strong land and a sturdy, and the plenteousest corner of the world, so rich a land that unneth it needeth help of any land, and every other land needeth help of England. England is full of mirth and of game, and men oft times able to mirth and game, free men of heart and with tongue, but the hand is more better and more free than the tongue.

Cedar is the name of the country in which dwelled the Ishmaelites, that were the children of Kedar, that was Ishmael's eldest son. And more truly they be there clept Agareni than Saraceni, though they mistake the name of Sarah in vain, and be proud thereof, as though they were gendered of Sarah. These men build no houses, but go about in large wildernesses, as wild men, and dwell in tents, and live by prey and by venison. Yet hereafter, as Methodius saith, they shall once be gathered together, and go out of the desert, and win and hold the roundness of the earth, eight weeks of years, and their way shall be called the way of anguish and of woe. For they shall overcome cities and kingdoms. And they shall slay priests in holy places, and lie there with women, and drink of holy vessels, and tie beasts to sepultures of holy saints, for the wickedness of the Christian men that shall be in that time. These and many other things he doth rehearse that Ishmaelites, men of Kedar, shall do in the world wide.

Ethiopia, blue men's land, had first that name of colour of men. For the sun is nigh, and roasteth and toasteth them. And so the colour of men showeth the strength of the star, for there is continual heat. For all that is under the south pole about the west is full of mountains, and about the middle full of gravel, and in the east side most desert and wilderness: and stretcheth from the west of Atlas toward the east unto the ends of Egypt, and is closed in the south with ocean, and in the north with the river Nile. In this land be many nations with divers faces wonderly and horribly shapen: Also therein be many wild beasts and serpents, and also Rhinoceros, and the beast that hight Cameleon, a beast with many colours. Also there be cockatrices and great dragons, and precious stones be taken out of their brains, Jacinth, and Chrysophrase, Topaz, and many other precious stones be found in those parts, and cinnamon is there gathered. There be two Ethiopias, one is in the east, and the other is in Mauritania in the west, and that is more near Spain. And then is Numidia, and the province of Carthage. Then is Getula, and at last against the course of the sun in the south is the land that hight Ethiopia adusta, burnt; and fables tell, that there beyond be the Antipodes, men that have their feet against our feet. The men of Ethiopia have their name of a black river, and that river is of the same kind as Nilus, for they breed reeds and bullrushes, and rise and wax in one time. In the wilderness there be many men wonderly shapen. Some oft curse the sun bitterly in his rising and downgoing, and they behold the sun and curse him always: for his heat grieveth them full sore. And other as Trogodites dig them dens and caves, and dwell in them instead of houses; and they eat serpents, and all that may be got; their noise is more fearful in sounding than the voice of other. Others there be which like beasts live without wedding, and dwell with women without law, and such be called Garamantes. Others go naked, and be not occupied with travail, and they be called Graphasantes. There be other that be called Bennii, and it is said, they have no heads, but they have eyes fixed in their breasts. And there be Satyrs, and they have only shape of men, and have no manners of mankind. Also in Ethiopia be many other wonders, there be Ethiops, saith Plinius, among whom all four-footed beasts be brought forth without ears, and also elephants. Also there be some that have a hound for their king, and divine by his moving, and do as they will. And other have three or four eyes in their foreheads, as it is said, not that it is so in kind, but that it is feigned, for they use principally looking and sight of arrows. Also some of them hunt lions and panthers, and live by their flesh, and their king hath only one eye in his forehead. Other men of Ethiopia live only by honeysuckles dried in smoke, and in the sun, and these live not past forty years.

In the over Egypt be many divers deserts, in whom are many monstrous and wonderful beasts. There be Pards, Tigers, Satyrs, Cockatrices, and horrible adders and serpents. For in the ends of Egypt and of Ethiopia fast by the well where men suppose is the head of Nilus that runneth by Egypt, be bred wild beasts, that hight Cacothephas, the which beast is little of body, and uncrafty of members and slow, and hath a full heavy head. And therefore they bear it always downward toward the earth, and that by ordinance of kind for the salvation of mankind, for it is so wicked and so venomous, that no man may behold it right in the face, but he die anon without remedy.

Fraunce hight Francia and Gallia also, and had first that name Francia of men of Germany, who were called Franci: and hath the Rhine and Germayn in the east side, and in the north-east side the mountains Alpes Pennini: and in the south the province of Narbonne, in the north-west the British ocean, and in the north the island of Britain.... This land of France is a rank country, and plentiful of trees, of vines, of corn, and of fruits, and is noble by the affluence of rivers and fountains; through the borders of which land run two most noble rivers, that is to wit, Rhone and Rhine. Therein be noble quarries and stones both to build and to rear buildings and houses upon, and therein be special manner stones, and namely in the ground about Paris, that is most passing, namely in a manner stone that is hight Gypsum, that men of that country call Plaster in their language, for the ground is glassy and bright, and by mineral virtue turneth into stone; this manner stone burnt and tempered with water, turneth into cement, and so thereof is made edifices and vaults, walls and diverse pavements. And such cement laid in works waxeth hard anon again as it were stone; and in France be many noble and famous cities, but among all Paris beareth the prize; for as sometime the city of Athens, mother of liberal arts and of letters, nurse of philosophers, and well of all sciences, made it solemn in science and in conditions among Greeks, so doth Paris in this time, not only France, but also all the other deal of Europe. For as mother of wisdom she receiveth all that cometh out of every country of the world, and helpeth them in all that they need, and ruleth all peaceably, and as a servant of soothness, she sheweth herself detty to wise men and unwise. This city is full good and mighty of riches, it rejoiceth in peace: there is good air of rivers according to philosophers, there be fair fields, meads, and mountains to refresh and comfort the eyen of them that be weary in study, there be convenable streets and houses, namely for studiers. And nevertheless the city is sufficient to receive and to feed all others that come thereto, and passeth all other cities in these things, and in such other like.

Though this province be little in space, yet it is wealthful of many special things and good. For this land is plenteous and full of pasture, of cattle, and of beasts, royal and rich of the best towns, havens of the sea, and of famous rivers, and well nigh all about is moisted with Scaldelia. The men thereof be seemly and fair of body and strong, and they get many children. And they be rich of all manner merchandises and chaffer, and generally fair and seemly of face, mild of will, and fair of speech, sad of bearing, honest of clothing, peaceable to their own neighbours, true and trusty to strangers, passing witty in wool craft, by their crafty working a great part of the world is succoured and holpen in woollen clothes. For of the principal wool which they have out of England, with their subtle craft be made many noble cloths, and be sent by sea and also by land into many diverse countries.

The men of Germany call men of this land Frisons, and between them and the Germans is great difference in clothing and in manner. For wellnigh all men be shorn round; and the more noble they be, the more worship they account to be shorn the more high. And the men be high of body, strong of virtue, stern and fierce of heart, and swift and quiver of body. And they use iron spears instead of arrows.... The men be free, and not subject to lordship of other nations, and put them in peril of death by cause of freedom. And they had liefer die than be under the yoke of thraldom. Therefore they forsake dignity of knighthood, and suffer none to rise and to be greater among them under the title of knighthood; but they be subject to Judges that they chose of themselves from year to year, which rule the community among them. They love well chastity, and punish all the unchaste right grievously:

And they keep their children chaste unto the time that they be of full age, and so when they be wedded, they get manly children and strong.

And, as it is said, some of the Indians till the earth, and some use chivalry, and some use merchandise and lead out chaffer; some rule and govern the community at best; and some be about the kings, and some be Justices and doomsmen, some give them principally to religions and to learning of wit and of wisdom. And as among all countries and lands India is the greatest and most rich: so among all lands India is most wonderful. For as Pliny saith, India aboundeth in wonders. In India be many huge beasts bred, and more greater hounds than in other lands. Also there be so high trees that men may not shoot to the top with an arrow, as it is said. And that maketh the plenty and fatness of the earth and temperateness of weather, of air, and of water. Fig trees spread there so broad, that many great companies of knights may sit at meat under the shadow of one tree. Also there be so great reeds and so long that every piece between two knots beareth sometime three men over the water. Also there be men of great stature, passing five cubits in height, and they never spit, nor have never headache nor toothache, nor sore eyes, nor they be not grieved with passing heat of the sun, but rather made more hard and sad therewith. Also their philosophers that they call Gymnosophists stand in most hot gravel from the morning till evening, and behold the sun without blemishing of their eyes. Also there, in some mountains be men with soles of the feet turned backwards, and the foot also with viii toes on one foot. Also there be some with hounds' heads, and be clothed in skins of wild beasts, and they bark as hounds, and speak none other wise: and they live by hunting and fowling: and they be armed with their nails and teeth, and be full many, about six score thousand as he saith. Also among some nations of India be women that bear never child but once, and the children wax whitehaired anon as they be born. There be satyrs and other men wondrously shapen. Also in the end of East India, about the rising of Ganges, be men without mouths, and they be clothed in moss and in rough hairy things, which they gather off trees, and live commonly by odour and smell at the nostrils. And they nother eat nother drink, but only smell odour of flowers and of wood apples, and live so, and they die anon in evil odour and smell. And other there be that live full long, and age never, but die as it were in middle age. Also some be hoar in youth, and black in age. Pliny rehearseth these wonders, and many other mo.

Yrlonde hight Hibernia, and is an island of the Ocean in Europe, and is nigh to the land of Britain, and is more narrow and straight than Britain, but it is more plenteous place.... In this land is much plenty of corn fields, of wells and of rivers, of fair meads and woods, of metal and of precious stones. For there is gendered a six cornered stone, that is to wit, Iris, that maketh a rainbow in the air, if it be set in the sun. And there is jet found, and white pearls. And concerning the wholesome air, Ireland is a good temperate country. There is little or none passing heat or cold; there be wonderful lakes, ponds, and wells. For there is a lake, in which if a staff or a pole of tree be pight, and tarrieth long time therein, the part that is in the earth turneth into iron, and the part that is in the water turneth into stone, and the part that is above the water, abideth still in its kind of tree. There is another lake in which in that thou throwest rods of hazel, it turneth those rods into ash: and ayen-ward if ye cast ashen rods therein, they turn into hazel. Therein be places in which dead carrions never rot: but abide there always uncorrupt. Also in Ireland is a little island, in which men die not, but when they be overcome with age, they be borne out of that island to die without. In Ireland is no serpent, no frogs, nor venomous addercop; but all the land is so contrary to venemous beasts that if the earth of that land be brought into another land, and spronge on the ground, it slayeth serpents and toads. Also venomous beasts flee Irish wool, skins, and fells. And if serpents or toads be brought into Ireland by shipping, they die anon.

Solinus speaketh of Ireland, and saith the inhabitants thereof be fierce, and lead an unhuman life. The people there use to harbour no guests, they be warriors, and drink men's blood that they slay, and wash first their faces therewith: right and unright they take for one.... Men of Ireland be singularly clothed and unseemly arrayed and scarcely fed, they be cruel of heart, fierce of cheer, angry of speech, and sharp. Nathless they be free hearted, and fair of speech and goodly to their own nation, and namely those men that dwell in woods, marshes, and mountains. These men be pleased with flesh, apples, and fruit for meat, and with milk for drink: and give them more to plays and to hunting, than to work and travail.

The land Scotia hath the name of Scots that dwell therein, and the same nation that was sometime first in Ireland, and all according thereto in tongue, in manners, and in kind. The men are light of heart, fierce, and courageous on their enemies. They love nigh as well death as thraldom, and they account it for sloth to die in bed, and a great worship and virtue to die in a field fighting against enemies. The men be of scarce living, and many suffer hunger long time, and eat selde tofore the sun going down, and use flesh, milk, meats, fish, and fruits more than Britons: and use to eat the less bread, and though the men be seemly enough of figure and of shape, and fair of face generally by kind, yet their own Scottish clothing disfigures them full much. And Scots be said in their own tongue of bodies painted, as it were cut and slit. For in old time they were marked with divers figures and shapes on their flesh and skin, made with iron pricks. And by cause of medlying with Englishmen, many of them have changed the old manners of Scots into better manners for the more part, but the wild Scots and Irish account great worship to follow their forefathers in clothing, in tongue, and in living, and in other manner doing. And despise somedeal the usages of other men in comparison to their own usage. And so each laboureth to be above, they detract and blame all other, and envy all other: they deride all other, and blame all other men's manners; they be not ashamed to lie, and they repute no man, of what nation, blood, or puissance so-ever he be, to be hardy and valiant, but themselves. They delight in their own; they love not peace. In that land is plenteous ground, merry woods, moist rivers and wells, many flocks of beasts. There be earth-tillers for quantity of the place enow.

Thanet is a little island of ocean, and is departed from Britain with a little arm of the sea, and hath wheat fields and noble grounds, and hath its name of death of serpents. For the earth of that land carried into any country of the world, slayeth serpents forthwith, as Isidore saith.

Finland is a country beside the mountains of Norway toward the east, and stretcheth upon the cliff of ocean: and is not full plenteous, but in wood, herbs, and grass. The men of that country be strange and somewhat wild and fierce: and they occupy themselves with witchcraft. And so to men that sail by their coasts, and also to men that abide with them for default of wind, they proffer wind to sailing, and so they sell wind. They use to make a clue of thread, and they make divers knots to be knit therein. And then they command to draw out of the clue unto three knots, or mo or less, as they will have the wind more soft or strong. And for their misbelief fiends move the air, and arise strong tempests or soft, as he draweth of the clue more or less knots. And sometimes they move the wind so strongly, that the wretches that believe in such doings, are drowned by rightful doom of God.

Iceland is the last region in Europe in the north beyond Norway. In the uttermost parts thereof it is always ice and frozen, and stretcheth upon the cliff of ocean toward the north, where the sea is frozen for great and strong cold. And Iceland hath the over Scythia in the east side, and Norway in the south, and the Irish ocean in the west, and the sea that is far in the north, and is called Iceland, as it were the land of ice and of glass. For it is said that there be mountains of snow froze as hard as ice or glass; there crystal is found. Also in that region are white bears most great and right fierce; that break ice and glass with their claws, and make many holes therein, and dive there-through into the sea, and take fish under the ice and glass, and draw them out through the same holes, and bring them to the cliff and live thereby. The land is barren, out-take a few places in the valleys, in the which places unneth grow oats. In the places that men dwell in, only grow herbs, grass, and trees. And in those places breed beasts, tame and wild. And so for the more part men of the land live by fish and by hunting of flesh. Sheep may not live there for cold. And therefore men of the land wear, for cold, fells and skins of bears and of wild beasts that they take with hunting. Other clothing may they not have, but it come of other lands. The men are full gross of body and strong and full white, and give them to fishing and hunting.

## V
# MEDIAEVAL NATURAL HISTORY--TREES

The seventeenth book of the "De Proprietatibus" deals with the properties of plants. The sources from which Bartholomew derives his information are Aristotle and Albertus Magnus' Gloss on the "De Vegetalibus," Albumazar, Pliny, Isaac on Foods, Hugo, and the Platearius. The text professes to deal with those trees and plants alone which are mentioned in the Gloss, but many others are incidentally mentioned, and we are thus enabled to learn the chief food-stuffs of our ancestors. The cereals of the time are wheat, barley, oats, and rye, just as at present; but the dinner-table of the day had neither turnip, cabbage, nor potato, and supplied their place with the parsnip, cole, and rape. Garlic, radishes, and lettuce were widely used, the former being valued in proportion to its power of overcoming any other odour. Flax seems to have been widely grown, and rushlights were then a luxury.

The subject of trees and plants does not so readily lend itself to fables as some other parts of natural history, but we refer the reader to the accounts of aloes, pepper, and mandragora as a specimen of the tales told, as our author says, "to make things dear, and of great price."

Aloes is a tree with good savour, and breedeth in India, and sometime a part thereof is set afire upon the altar in the stead of incense. It is found in the great river of Babylon, that joineth with a river of Paradise. Therefore many men trow that the aforesaid tree groweth among the trees of Paradise, and cometh out of Paradise by some hap or drift into [the] river of Ind. Men that dwell by that river take this tree out of the water by nets, and keep it to the use of medicine, for it is a good medicinal tree.

Of Cannel and of Cassia men told fables in old time, that it is found in birds' nests, and specially in the Phoenix' nest. And may not be found, but what falleth by its own weight, or is smitten down with lead arrows. But these men do feign, to make things dear and of great price; but as the sooth meaneth, cannel groweth among the Trogodites in the little Ethiopia, and cometh by long space of the sea in ships to the haven of Gelenites. No man hath leave to gather thereof tofore the sun-rising, nor after the sun going down. And when it is gathered, the priest by measure dealeth the branches and taketh thereof a part; and so by space of time, merchants buy that other deal.

Of this tree [Bays] speaketh the Master in History, and saith that Rebecca (Gen. xvii.) for trembling of nations she had seen in them that perished, laid a manner laurel tree that she called Tripodem under her head, and sat her upon boughs of an herb that hight Agnus Castus, for to use very revelations and sights and not fantasies.

The Emperor Tiberius Caesar in thundering and lightning used a garland of Laurel Tree on his head against dread of lightning, as it is said. Also Plinius telleth a wonder thing, that the emperor sat by Drusilla the empress in a certain garden, and an eagle threw from a right high place a wonder white hen into the empress' lap whole and sound, and the hen held in her bill a bough of laurel tree full of bays, and Diviners took heed to the hen, and sowed the bays, and kept them wisely, and of them came a wood, that was called Silva Triumphans, as it were the wood of worship for victory and mastery.

The green leaves thereof, that smell full well if they be stamped, heal stinging of bees and of wasps, and do away all swellings, and keep books and clothes there it is among from moths and other worms, and save them fro fretting and gnawing. The fruit of laurel trees are called bays, and are brown or red without, and white within and unctuous.

It is said that a hind taught first the virtue of diptannus, for she eateth this herb that she may calve easilier and sooner; and if she be hurt with an arrow, she seeketh this herb and eateth it, which putteth the iron out of the wound.

And ash hath so great virtue that serpents come not in shadow thereof in the morning nor at even. And if a serpent be set within a fire and ash leaves, he will flee into the fire sooner than into the leaves.

Beans be damned by Pythagoras' sentence, for it is said, that by oft use thereof the wits are dulled and cause many dreams. Or else as other men mean, for dead men's souls be therein. Therefore Varro saith that the bishop should not eat beans. And many medley beans with bread corn, to make the bread more heavy.

The stalk [of wheat] is called Stipula as ustipula, and hath that name of usta, burnt. For when it is gathered some of the straw is burnt to help and amend the land. And some is kept to fodder of beasts, and is called Palea: for it is first meat that is laid tofore beasts, namely in some countries as in Tuscany. As Pliny saith, if the seed be touched with tallow or grease it is spoilt and lost. Among the best wheat sometimes grow ill weeds and venomous, as cockle and other such, also there it is said, of corrupt dew that cleaveth to the leaves cometh corruption in corn, and maketh it as it were red or rusty. Among all manner corn, wheat beareth the prize, and to mankind nothing is more friendly, nothing more nourishing.

Flax groweth in even stalks, and bears yellow flowers or blue, and after cometh hops, and therein is the seed, and when the hop beginneth to wax, then the flax is drawn up and gathered all whole, and is then lined, and afterward made to knots and little bundles, and so laid in water, and lieth there long time. And then it is taken out of the water, and laid abroad till it be dried, and twined and wend in the sun, and then bound in pretty niches and bundles. And afterward knocked, beaten, and brayed, and carfled, rodded and gnodded, ribbed and heckled, and at the last spun. Then the thread is sod and bleached, and bucked, and oft laid to drying, wetted and washed, and sprinkled with water until that it be white, after divers working and travail.

Flax is needful to divers uses. For thereof is made clothing to wear, and sails to sail, and nets to fish and to hunt, and thread to sew, ropes to bind, and strings to shoot, bonds to bind, lines to mete and to measure, and sheets to rest in, and sacks, bags, and purses, to put and to keep things in. And so none herb is so needful, to so many divers uses to mankind, as is the flax.

Ryndes thereof [i.e. of Mandragora] sodden in wine cause sleep, and abate all manner of soreness, and so that time a man feeleth unneth though he be cut, but yet Mandragora must be warily used: for it slayeth if men take much thereof.... They that dig Mandragora be busy to beware of contrary winds while they dig, and make three circles about with a sword, and abide with the digging unto the sun going down, and trow so to have the herb with the chief virtues.

Papyrus is a manner rush, that is dried to kindle fire and lanterns, and hight the feeding of fire. And this herb is put to burn in prickets and in tapers. The rind is stripped off unto the pith, and is so dried, and a little is left of the rind on the one side, to sustain the tender pith; and the less is left of the rind, the more clear the pith burneth in a lamp, and is the sooner kindled. And about Memphis and in Ind be such great rushes, that they make boats thereof, as the Gloss saith. And Alexander's Story saith the same.

And of rushes are charters made, in the which were epistles written, and sent by messengers. Also of rushes be made paniers, boxes, and cases, and baskets to keep letters and other things in. And also they make thereof paper to write with.

Pepper is the seed or the fruit of a tree that groweth in the south side of the hill Caucasus, in the strong heat of the sun. And serpents keep the woods that pepper groweth in. And when the woods of pepper are ripe, men of that country set them on fire, and chase away the serpents by violence of fire. And by such burning the grain of pepper that was white by kind, is made black and rively.

Woods be wild places, waste and desolate, that many trees grow in without fruit, and also few having fruit. In these woods be oft wild beasts and fowls, therein grow herbs, grass leas, and pasture, and namely medicinal herbs in woods be found. In summer woods are beautied with boughs and branches, with herbs and grass. In woods is place of deceit and hunting. For therein wild beasts are hunted, and watches and deceits are ordained and set of hounds and of hunters. There is place of hiding and of lurking, for oft in woods thieves are hid, and oft in their awaits and deceits passing men come, and are spoiled and robbed, and oft slain. And so for many and divers ways and uncertain, strange men oft err and go out of the way, and take uncertain ways, and the way that is unknown tofore the way that is known, and come oft to the place there thieves lie in await, and not without peril. Therefore be oft knots made on trees and in bushes, in boughs and in branches of trees, in token and mark of the highway, to show the certain and sure way to wayfaring men; but oft the thieves in turning and meeting of ways, change such knots and signs, and beguile many men, and bring them out of the right way by false tokens and signs.

It hath many hard twigs and branches with knots, and therewith often children are chastised and beaten on the bare buttocks and loins. And of the boughs and branches thereof are besoms made to sweep and to clean houses of dust and of other uncleanness. Wild men of woods and forests use that seed in stead of bread. And this tree hath much sour juice, and somewhat biting. And men use therefore in springing time and in harvest to slit the rinds, and to gather the humour that cometh out thereof, and drink it in stead of wine.

Hards is the cleansing of hemp or of flax. For with much breaking, heckling, and rubbing, hards are departed fro the substance of hemp and of flax, and is great when it is departed, and more knotty, short, and rough. And is therefore not full able to be spun for thread thereof to be made, nathless thereof is thread spun that is full great, uneven, and full of knobs, and thereof are made bonds and bindings, and matches or candles; for it is full dry and taketh soon fire and burneth.

A board hight table, and is areared and set upon feet, and compassed with a list about. And, in another manner, table is a playing board, that men play on at the dice and other games; and this manner of table is double, and arrayed with divers colours. In the third manner it is a thin plank and plane, and therein are letters writ with colours, and sometimes small shingles are planed and made somedeal hollow in either side, and filled full of wax, black, green, or red, to write therein.

Boards and tables garnish houses, nathless when they be set in solar floors, they serve all men and beasts that are therein. Then they be dressed, hewed, and planed, and made convenable to use of ships, of bridges, of hulks, and coffers, and many other needful things of building. Also in shipbreach men flee to a board, and are oft saved in peril.

Roofs are trees areared and stretched fro the walls up to the top of the house, and bear up the covering thereof. And stand wide beneath, and come together upwards, and so they nigh nearer and nearer, and are joined either to other in the top of the house. It holdeth up heling, slates, shingle, and laths. The lath is long and somewhat broad, and plain and thin, and is nailed thwart over to the rafters, and thereon hang slates, tiles, and shingles. The rafters are strong and square, and hewn plain And are made fair within with fair joists and boards.

A vineyard is busily tilthed and kept, and purged and cleaned of superfluities, and oft visited and overseen of the earth tilthers and keepers of vines, that it be not apaired neither destroyed with beasts, and is closed about with walls and with hedges, and a wait is there set in a high place to keep the vineyard that the fruit be not destroyed. And is left in winter without keeper or waiter, but in harvest time many come and haunt the vineyard. In winter the vineyard is full pale, and waxeth green and bloometh in springing time and in summer, and smelleth full sweet, and is pleasant with fruit in harvest time. The smell of the vineyard that bloometh is contrary to all venomous things, and therefore when the vineyard bloometh, adders and serpents flee, and toads also, and may not sustain and suffer the noble savour thereof.

Foxes lurk and hide themselves under vine leaves, and gnaw covetously and fret the grapes of the vineyard, and namely when the keepers and wards be negligent and reckless, and it profiteth not that some unwise men do, that close within the vineyard hounds, that are adversaries to foxes. For few hounds, so closed, waste and destroy more grapes than many foxes should destroy that come and eat thereof thievishly. Therefore wise wardens of vineyards be full busy to keep, that no swine nor tame hounds nor foxes come in to the vineyard. From fretting and gnawing of flies and of other worms, a vineyard may not be kept nor saved, but by His succour and help that all thing hath and pursueth in His power and might, and keepeth and saveth all lordly and mighty.

The worthiness and praising of wine might not Bacchus himself describe at the full, though he were alive. For among all liquors and juice of trees, wine beareth the prize, for passing all liquors, wine moderately drunk most comforteth the body, and gladdeth the heart, and saveth wounds and evils. Wine strengtheneth all the members of the body, and giveth to each might and strength, and deed and working of the soul showeth and declareth the goodness of wine. And wine breedeth in the soul forgetting of anguish, of sorrow, and of woe, and suffereth not the soul to feel anguish and woe. Wine sharpeth the wit and maketh it cunning to enquire things that are hard and subtle, and maketh the soul bold and hardy, and so the passing nobility of wine is known. And use of wine accordeth to all men's ages and times and countries, if it be taken in due manner, and as his disposition asketh that drinketh it.

Red wine that is temperate in its qualities, and is drunk temperately and in due manner, helpeth kind and gendreth good blood, and maketh savour in meat and in drink, and exciteth desire and appetite, and comforteth the virtue of life and of kind, and helpeth the stomach to have appetite, and to have and to make good digestion. And quencheth thirst, and changeth the passions of the soul and thoughts out of evil into good. For it turneth the soul out of cruelness into mildness, out of covetousness into largeness, out of pride into meekness, and out of dread into boldness. And shortly to speak, wine drunk measurably is health of body and of soul.

And nothing is worse passing out of measure. And so Andronides, a clear man of wit and of wisdom, wrote to the great Alexander, to restrain wine kind in drinking, and said in this manner:--"King, have mind that thou drinkest blood of the earth, for wine drinking untemperately is to mankind heavy and venomous." And if Alexander had done by his counsel, truly he had not slain his own friend in drunkenness. If wine be often taken, anon by drunkenness it quencheth the sight of reason, and comforteth beastly madness, and so the body abideth as it were a ship in the sea without stern and without lodesman, and as chivalry without prince or duke.

## VI
## MEDIAEVAL NATURAL HISTORY--BIRDS AND FISHES

In following out his plan of describing the productions of each element before considering the next in order, Bartholomew was led to consider air and its products early in his scheme. Accordingly his twelfth book is devoted to birds, and his thirteenth to the inhabitants of the waters. There is hardly any reason in these books for omitting any part more than another except space, but the editor hopes that those chosen will put the reader in possession of a key to the more common allusions in pre-Restoration literature.

When the editor spoke of the wholesale way in which our author is conveyed by Elizabethan poets, he had in mind this and the following chapters. A single example will show this. Let the reader compare the account of the peacock with the following stanza from Chester's "Love's Martyr":

"The proud sun-braving peacocke with his feathers,
Walkes all along, thinking himself a king,
And with his voice prognosticates all weathers,
Although, God knows, but badly he doth sing;
But when he looks downe to his base blacke feete,
He droopes and is asham'd of things unmeet."

Our author's knowledge of birds is largely derived--the authentic from Aristotle; the legendary from the Fathers, Ambrose, Austin, Basil, and Gregory,--the Gloss,--and from Pliny. Some of these legends seem to be pointed at in the Hebrew

Scriptures. Thus Ps. ciii. 5, "Thy youth is renewed like the eagle's," either gave rise to, or refers to, the tradition quoted in our account of the eagle: and likewise Job xxxviii. 41, and Ps. cxlvii. 9, seem to be responsible for the tradition in the account of the raven. It would be interesting to learn whether any independent traditions of this nature exist.

It is worth pointing out that our author has contributed to the "Gesta Romanorum" several stories. The "wild tale," as Warton calls it, of the elephant and the maidens, as well as the story of "the storke wreker of avouterie" mentioned by Chaucer in the "Assemblie of Foules," and derived from Neckham, and the similar tale of the lioness, obtained their wide circulation through the popularity of Bartholomew's book. It would be an interesting task to trace these tales to their origin, but this is neither the place nor the time to do so; and the editor similarly leaves to lovers of Shakespeare the pleasure of proving to themselves his intimate acquaintance with the book.

In the part of the chapter quoted from the thirteenth book, the editor has tried to get together some of those stories which impressed people's minds most. Such a one is the tale of the remora. We remember Jonson's use of it in the "Poetaster":

"Death, I am seized here
By a land remora; I cannot stir
Nor move, but as he pleases."

Other tales remind us of Olaus Magnus, and some of them are plainly Eastern.

Now it pertaineth to speak of birds and fowls, and in particular and first of the eagle, which hath principality among fowls. Among all manner kinds of divers fowls, the eagle is the more liberal and free of heart. For the prey that she taketh, but it be for great hunger, she eateth not alone, but putteth it forth in common to fowls that follow her. But first she taketh her own portion and part. And therefore oft other fowls follow the eagle for hope and trust to have some part of her prey. But when the prey that is taken is not sufficient to herself, then as a king that taketh heed to a community, she taketh the bird that is next to her, and giveth it among the others, and serveth them therewith.

Austin saith, and Plinius also, that in age the eagle hath darkness and dimness in eyen, and heaviness in wings. And against this disadvantage she is taught by kind to seek a well of springing water, and then she flieth up into the air as far as she may, till she be full hot by heat of the air, and by travail of flight, and so then by heat the pores are opened and the feathers chafed, and she falleth suddenly in to the well, and there the feathers are changed, and the dimness of her eyes is wiped away and purged, and she taketh again her might and strength.

The eagle's feathers done and set among feathers of wings of other birds corrupteth and fretteth them. As strings made of wolf-gut done and put into a lute or in an harp among strings made of sheep-gut do destroy, and fret, and corrupt the strings made of sheep-gut, if it so be that they be set among them, as in a lute or in an harp, as Pliny saith.

Among all fowls, in the eagle the virtue of sight is most mighty and strong. For in the eagle the spirit of sight is most temperate and most sharp in act and deed of seeing and beholding the sun in the roundness of its circle without blemishing of eyen. And the sharpness of her sight is not rebounded again with clearness of light of the sun, nother disperpled. There is one manner eagle that is full sharp of sight, and she taketh her own birds in her claws, and maketh them to look even on the sun, and that ere their wings be full grown, and except they look stiffly and steadfastly against the sun, she beateth them, and setteth them even tofore the sun. And if any eye of any of her birds watereth in looking on the sun she slayeth him, as though he went out of kind, or else driveth him out of the nest and despiseth him, and setteth not by him.

The goshawk is a royal fowl, and is armed more with boldness than with claws, and as much as kind taketh from her in quantity of body, it rewardeth her with boldness of heart. And two kinds there be of such fowls, for some are tame and some are wild. And she that is tame taketh wild fowls and taketh them to her own lord, and she that is wild taketh tame fowls. And this hawk is of a disdainful kind. For if she fail by any hap of the prey that she reseth to, that day unneth she cometh unto her lord's hand. And she must have ordinate diet, nother too scarce, ne too full. For by too much meat she waxeth ramaious or slow, and disdaineth to come to reclaim. And if the meat be too scarce then she faileth, and is feeble and unmighty to take her prey. Also the eyen of such birds should oft be seled and closed, or hid, that she bate not too oft from his hand that beareth her, when she seeth a bird that she desireth to take; and also her legs must be fastened with gesses, that she shall not fly freely to every bird. And they be borne on the left hand, that they may somewhat take of the right hand, and be fed therewith.

And so such tame hawks be kept in mews, that they may be discharged of old feathers and hard, and be so renewed in fairness of youth. Also men give them meat of some manner of flesh, which is some-deal venomous, that they may the sooner change their feathers. And smoke grieveth such hawks and doth them harm. And therefore their mews must be far from smoky places, that their bodies be not grieved with bitterness of smoke, nor their feathers infect with blackness of smoke. They should be fed with fresh flesh and bloody, and men should use to give them to eat the hearts of fowls that they take. All the while they are alive and are strong and mighty to take their prey, they are beloved of their lords, and borne on hands, and set on perches, and stroked on the breast and on the tail, and made plain and smooth, and are nourished with great business and diligence. But when they are dead, all men hold them unprofitable and nothing worth, and be not eaten, but rather thrown out on dunghills.

The properties of bees are wonderful noble and worthy. For bees have one common kind as children, and dwell in one habitation, and are closed within one gate: one travail is common to them all, one meat is common to them all, one common working, one common use, one fruit and flight is common to them all, and one generation is common to them all.

Also maidenhood of body without wem is common to them all, and so is birth also. For they are not medlied with service of Venus, nother resolved with lechery, nother bruised with sorrow of birth of children. And yet they bring forth most swarms of children.

Bees make among them a king, and ordain among them common people. And though they be put and set under a king, yet they are free and love their king that they make, by kind love, and defend him with full great defence, and hold [it] honour and worship to perish and be spilt for their king, and do their king so great worship that none of them dare go out of their house, nor to get meat, but if the king pass out and take the principality of flight. And bees chose to their king him that is most worthy and noble in highness and fairness, and most clear in mildness, for that is chief virtue in a king. For though their king have a sting yet he useth it not in wreck. And also bees that are unobedient to the king, they deem themselves by their own doom for to die by the wound of their own sting. And of a swarm of bees is none idle. Some fight, as it were in battle, in the field against other bees, some are busy about meat, and some watch the coming of showers. And some behold concourse and meting of dues, and some make wax of flowers, and some make cells now round, now square with wonder binding and joining, and evenness. And yet nevertheless, among so diverse works none of them doth espy nor wait to take out of other's travail, neither taketh wrongfully, neither stealeth meat, but each seeketh and gathereth by his own flight and travail among herbs and flowers that are good and convenable.

Bees sit not on fruit but on flowers, not withered but fresh and new, and gather matter of the which they make both honey and wax. And when the flowers that are nigh unto them be spent, then they send spies for to espy meat in further places. And if the night falleth upon them in their journey, then they lie upright to defend their wings from rain, and from dew, that they may in the morrow tide fly the more swifter to their work with their wings dry and able to fly. And they ordain watches after the manner of castles, and rest all night until it be day, till one bee wake them all with twice buzzing or thrice, or with some manner trumping; then they fly all, if the day be fair on the morrow. And the bees that bring and bear what is needful, dread blasts of wind, and fly therefore low by the ground when they be charged, lest they be letted with some manner of blasts, and charge themselves sometimes with gravel or with small stones, that they may be the more stedfast against blasts of wind by heaviness of the stones.

The obedience of bees is wonderful about the king, for when he passeth forth, all the swarm in one cluster passeth with him. And he is beclipped about with the swarm, as it were with an host of knights. And is then unneth seen that time for the multitude that followeth and serveth him, and when the people of bees are in travail, he is within, and as it were governor, and goeth about to comfort others for to work. And only he is not bound to travail. And all about him are certain bees with stings, as it were champions, and continual wardens of the king's body. And he passeth selde out, but when all the swarm shall go out. His outgoing is known certain days tofore by voice of the host, as it were arraying itself to pass out with the king.

The culvour is messager of peace, ensample of simpleness, clean of kind, plenteous in children, follower of meekness, friend of company, forgetter of wrongs. The culvour is forgetful. And therefore when the birds are borne away, she forgetteth her harm and damage, and leaveth not therefore to build and breed in the same place. Also she is nicely curious. For sitting on a tree, she beholdeth and looketh all about toward what part she will fly, and bendeth her neck all about as it were taking avisement. But oft while she taketh avisement of flight, ere she taketh her flight, an arrow flieth through her body, and therefore she faileth of her purpose, as Gregory saith.

Also as Ambrose saith, in Egypt and in Syria a culvour is taught to bear letters, and to be messager out of one province into another. For it loveth kindly the place and the dwelling where it was first fed and nourished. And be it never so far borne into far countries, always it will return home again, if it be restored to freedom. And oft to such a culvour a letter is craftily bound under the one wing, and then it is let go. Then it flieth up into the air, and ceaseth never till it come to the first place in which it was bred. And sometimes in the way enemies know thereof, and let it with an arrow, and so for the letters that it beareth, it is wounded and slain, and so it beareth no letter without peril. For oft the letter that is so borne is cause and occasion of the death of it.

The crow is a bird of long life, and diviners tell that she taketh heed of spyings and awaitings, and teacheth and sheweth ways, and warneth what shall fall. But it is full unlawful to believe, that God sheweth His privy counsel to crows. It is said that crows rule and lead storks, and come about them as it were in routs, and fly about the storks and defend them, and fight against other birds and fowls that hate storks. And take upon them the battle of other birds, upon their own peril. And an open proof thereof is: for in that time, that the storks pass out of the country, crows are not seen in places there they were wont to be. And also for they come again with sore wounds, and with voice of blood, that is well known, and with other signs and tokens and show that they have been in strong fighting. Also there it is said, that the mildness of the bird is wonderful. For when father and mother in age are both naked and bare of covering of feathers, then the young crows hide and cover them with their feathers, and gather meat and feed them.

The raven beholdeth the mouths of her birds when they yawn. But she giveth them no meat ere she know and see the likeness of her own blackness, and of her own colour and feathers. And when they begin to wax black, then afterward she feedeth them with all her might and strength. It is said that ravens' birds are fed with dew of heaven all the time that they have no black feathers by benefit of age. Among fowls, only the raven hath four and sixty changings of voice.

The swan feigneth sweetness of sweet songs with accord of voice, and he singeth sweetly for he hath a long neck diversely bent to make divers notes. And it is said that, in the countries that are called Hyperborean, the harpers harping before, the swans' birds fly out of their nests and sing full merrily. Shipmen trow that it tokeneth good if they meet swans in peril of shipwreck. Always the swan is the most merriest bird in divinations. Shipmen desire this bird for he dippeth not down in the waves. When the swan is in love he seeketh the female, and pleaseth her with beclipping of the neck, and draweth her to him- ward; and he joineth his neck to the female's neck, as it were binding the necks together.

Phoenix is a bird, and there is but one of that kind in all the wide world. There- fore lewd men wonder thereof, and among the Arabs, there this bird is bred, he is called singular--alone. The philosopher speaketh of this bird and saith that phoenix is a bird without make, and liveth three hundred or five hundred years: when the which years are past, and he feeleth his own default and feebleness, he maketh a nest of right sweet-smelling sticks, that are full dry, and in summer when the western wind blows, the sticks and the nest are set on fire with burning heat of the sun, and burn strongly. Then this bird phoenix cometh willfully into the burning nest, and is there burnt to ashes among these burning sticks, and within three days a little worm is gendered of the ashes, and waxeth little and little, and taketh feath- ers and is shapen and turned to a bird. Ambrose saith the same in the Hexameron: Of the humours or ashes of phoenix ariseth a new bird and waxeth, and in space of time he is clothed with feathers and wings and restored into the kind of a bird, and is the most fairest bird that is, most like to the peacock in feathers, and loveth the wilderness, and gathereth his meat of clean grains and fruits. Alan speaketh of this bird and saith, that when the highest bishop Onyas built a temple in the city of Heliopolis in Egypt, to the likeness of the temple in Jerusalem, on the first day of Easter, when he had gathered much sweet-smelling wood, and set it on fire upon the altar to offer sacrifice, to all men's sight such a bird came suddenly, and fell into the middle of the fire, and was burnt anon to ashes in the fire of the sacrifice, and the ashes abode there, and were busily kept and saved by the commandments of the priests, and within three days, of these ashes was bred a little worm, that took the shape of a bird at the last, and flew into the wilderness.

The crane is a bird of great wings and strong flight, and flieth high into the air to see the countries towards the which he will draw. And is a bird that loveth birds of his own kind, and they living in company together have a king among them and fly in order. And the leader of the company compelleth the company to fly aright, crying as it were blaming with his voice. And if it hap that he wax hoarse, then another crane cometh after him, and taketh the same office. And after they fall to the earth crying, for to rest, and when they sit on the ground, to keep and save them, they ordain watches that they may rest the more surely, and the wakers stand upon one foot, and each of them holdeth a little stone in the other foot, high from the earth, that they may be waked by falling of the stone, if it hap that they sleep.

A griffin is accounted among flying things (Deut. xiiii.) and there the Gloss saith, that the griffin is four-footed, and like to the eagle in head and in wings, and is like to the lion in the other parts of the body. And dwelleth in those hills that are called Hyperborean, and are most enemies to horses and men, and grieveth them most, and layeth in his nest a stone that hight Smaragdus against venomous beasts of the mountain.

A pelican is a bird of Egypt, and dwelleth in deserts beside the river Nile. All that the pelican eateth, he plungeth in water with his foot, and when he hath so plunged it in water, he putteth it into his mouth with his own foot, as it were with an hand. Only the pelican and the popinjay among fowls use the foot instead of an hand.

The pelican loveth too much her children. For when the children be haught, and begin to wax hoar, they smite the father and the mother in the face, wherefore the mother smiteth them again and slayeth them. And the third day, the mother smiteth herself in her side, that the blood runneth out, and sheddeth that hot blood on the bodies of her children. And by virtue of that blood, the birds that were before dead quicken again.

Master Jacobus de Vitriaco in his book of the wonders of the Eastern parts telleth another cause of the death of pelicans' birds. He saith that the serpent hateth kindly this bird. Wherefore when the mother passeth out of the nest to get meat, the serpent climbeth on the tree, and stingeth and infecteth the birds. And when the mother cometh again, she maketh sorrow three days for her birds, as it is said. Then (he saith) she smiteth herself in the breast and springeth blood upon them, and reareth them from death to life, and then for great bleeding the mother waxeth feeble, and the birds are compelled to pass out of the nest to get themselves meat. And some of them for kind love feed the mother that is feeble, and some are unkind and care not for the mother, and the mother taketh good heed thereto, and when she cometh to her strength, she nourisheth and loveth those birds that fed her in her need, and putteth away her other birds, as unworthy and unkind, and suffereth them not to dwell nor live with her.

The peacock hath an unsteadfast and evil shapen head, as it were the head of a serpent, and with a crest. And he hath a simple pace, and small neck and areared, and a blue breast, and a tail full of eyes distinguished and high with wonder fairness, and he hath foulest feet and rivelled. And he wondereth of the fairness of his feathers, and areareth them up as it were a circle about his head, and then he looketh to his feet, and seeth the foulness of his feet, and like as he were ashamed he letteth his feathers fall suddenly, and all the tail downward, as though he took no heed of the fairness of his feathers. And as one saith, he hath the voice of a fiend, head of a serpent, pace of a thief. For he hath an horrible voice.

In this bird [the vulture] the wit of smelling is best. And therefore by smelling he savoureth carrions that be far from him, that is beyond the sea, and ayenward. Therefore the vulture followeth the host that he may feed himself with carrions of men and of horses. And therefore (as a Diviner saith), when many vultures come and fly together, it tokeneth battle. And they know that such a battle shall be, by some privy wit of kind. He eateth raw flesh, and therefore he fighteth against other fowls because of meat, and he hunteth fro midday to night, and resteth still fro the sunrising to that time. And when he ageth, his over bill waxeth long and crooked over the nether, and [he] dieth at the last for hunger.

And some men say, by error of old time, that the vulture was sometime a man, and was cruel to some pilgrims, and therefore he hath such pain of his bill, and dieth for hunger, but that is not lawful to believe.

Jorath saith, that there is a great fish in the sea, that hight Bellua, that casteth out water at his jaws with vapour of good smell, and other fish feel the smell and follow him, and enter and come in at his jaws following the smell, and he swalloweth them and is so fed with them. Also he saith that Dolphins know by the smell if a dead man, that is on the sea, ate ever of Dolphin's kind; and if the dead man hath eat thereof, he eateth him anon; and if he did not, he keepeth and defendeth him fro eating and biting of other fish, and shoveth him, and bringeth him to the cliff with his own working?

Enchirius is a little fish unneth half a foot long: for though he be full little of body, nathless he is most of virtue. For he cleaveth to the ship, and holdeth it still stedfastly in the sea, as though the ship were on ground therein. Though winds blow, and waves arise strongly, and wood storms, that ship may not move nother pass. And that fish holdeth not still the ship by no craft, but only cleaving to the ship. It is said of the same fish that when he knoweth and feeleth that tempests of wind and weather be great, he cometh and taketh a great stone, and holdeth him fast thereby, as it were by an anchor, lest he be smitten away and thrown about by waves of the sea. And shipmen see this and beware that they be not overset unwarily with tempest and with storms.

The crab is enemy to the oyster. For he liveth by fish thereof with a wonderful wit. For because that he may not open the hard shell of the oyster, he spieth and awaiteth when the oyster openeth, and then the crab, that lieth in await, taketh a little stone, and putteth it between the shells, that the oyster may not close himself. And when the closing is so let, the crab eateth and gnaweth the flesh of the oyster.

It is said that the whale hath great plenty of sperm, and after that he gendereth, superfluity thereof fleeteth above the water; and if it be gathered and dried it turneth to the substance of amber. And in age, for greatness of body, on his ridge powder and earth is gathered, and so digged together that herbs and small trees and bushes grow thereon, so that that great fish seemeth an island. And if shipmen come unwarily thereby, unneth they scape without peril. For he throweth as much water out of his mouth upon the ship, that he overturneth it sometime or drowneth it.

Also he is so fat that when he is smitten with fishers' darts he feeleth not the wound, but it passeth throughout the fatness. But when the inner fish is wounded, then is he most easily taken. For he may not suffer the bitterness of the salt water, and therefore he draweth to the shoreward. And also he is so huge in quantity, that when he is taken, all the country is better for the taking. Also he loveth his whelps with a wonder love, and leadeth them about in the sea long time. And if it happeth that his whelps be let with heaps of gravel, and by default of water, he taketh much water in his mouth, and throweth upon them, and delivereth them in that wise out of peril, and bringeth them again into the deep sea. And for to defend them he putteth himself against all things that he meeteth if it be noyful to them, and setteth them always between himself and the sun on the more secure side. And when strong tempest ariseth, while his whelps are tender and young, he swalloweth them up into his own womb. And when the tempest is gone and fair weather come, then he casteth them up whole and sound.

Also Jorath saith, that against the whale fighteth a fish of serpent's kind, and is venomous as a crocodile. And then other fish come to the whale's tail, and if the whale be overcome the other fish die. And if the venomous fish may not overcome the whale, then he throweth out of his jaws the whale throweth out of his mouth a sweet smelling smoke, and putteth off the stinking smell, and defendeth and saveth himself and his in that manner wise.

# VII
# MEDIAEVAL NATURAL HISTORY--ANIMALS

The eighteenth book of the "De Proprietatibus" is devoted to the properties of animals. It is composed of selections from Pliny and Aristotle, from the works of the mediaeval physicians and romancers, from Magister Jacobus de Vitriaco, from the "Historia Alexandri Magni de Proeliis," from Physiologus and the Bestiarium.

The editor has been obliged to reduce some of these extracts to make room for others. Among these the reader will find many examples of those legends, which made up the popular Natural History of early days, originally imported from the East through Spain and Italy. The memory of these survives even now in our popular locutions. "Licked into shape" refers to the tale we give in our account of the bear. The royal nature of the lion is a commonplace: Jonson and Spenser speak of the sweet breath of the panther. Drayton, in his "Heroical Epistles," quotes the siren and the hyena as examples:

> "To call for aid, and then to lie in wait,
> So the hyena murthers by deceit,
> By sweet enticement sudden death to bring,
> So from the rocks th' alluring mermaids sing."

Trevisa has invented an adjective for us that expresses the midnight caterwaul--"ghastful." Bartholomew probably suffered from those two minor curses of humanity--the amorous cat and the wandering cur. But he has preserved for us a noble eulogy of the dog, and has a reference to the tale of the dog of Montargis, the standing example of canine fidelity among a chivalrous folk.

It is said, that in India is a beast wonderly shapen, and is like to the bear in body and in hair, and to a man in face. And hath a right red head, and a full great mouth, and an horrible, and in either jaw three rows of teeth distinguished atween. The outer limbs thereof be as it were the outer limbs of a lion, and his tail is like to a wild scorpion, with a sting, and smiteth with hard bristle pricks as a wild swine, and hath an horrible voice, as the voice of a trumpet, and he runneth full swiftly, and eateth men. And among all beasts of the earth is none found more cruel, nor more wonderly shape, as Avicenna saith. And this beast is called Baricos in Greek.

The boar is so fierce a beast, and also so cruel, that for his fierceness and his cruelness, he despiseth and setteth nought by death, and he reseth full piteously against the point of a spear of the hunter. And though it be so that he be smitten or sticked with a spear through the body, yet for the greater ire and cruelness in heart that he hath, he reseth on his enemy, and taketh comfort and heart and strength for to wreak himself on his adversary with his tusks, and putteth himself in peril of death with a wonder fierceness against the weapon of his enemy, and hath in his mouth two crooked tusks right strong and sharp, and breaketh and rendeth cruelly with them those which he withstandeth. And useth the tusks instead of a sword. And hath a hard shield, broad and thick in the right side, and putteth that always against his weapon that pursueth him, and useth that brawn instead of a shield to defend himself. And when he spieth peril that should befall, he whetteth his tusks and frotteth them, and assayeth in that while fretting against trees, if the points of his tusks be all blunt. And if he feel that they be blunt, he seeketh a herb which is called Origanum, and gnaweth it and cheweth it, and cleanseth and comforteth the roots of his teeth therewith by vertue thereof.

The ass is fair of shape and of disposition while he is young and tender, or he pass into age. For the elder the ass is, the fouler he waxeth from day to day, and hairy and rough, and is a melancholy beast, that is cold and dry, and is therefore kindly heavy and slow, and unlusty, dull and witless and forgetful. Nathless he beareth burdens, and may away with travail and thraldom, and useth vile meat and little, and gathereth his meat among briars and thorns and thistles.... And the ass hath another wretched condition known to nigh all men. For he is put to travail over-night, and is beaten with staves, and sticked and pricked with pricks, and his mouth is wrung with a bernacle, and is led hither and thither, and withdrawn from leas and pasture that is in his way oft by the refraining of the bernacle, and dieth at last after vain travails, and hath no reward after his death for the service and travail that he had living, not so much that his own skin is left with him, but it is taken away, and the carrion is thrown out without sepulture or burials; but it be so much of the carrion that by eating and devouring is sometimes buried in the wombs of hounds and wolves.

And such [adders] lie in await for them that sleep: and if they find the mouth open of them or of other beasts, then they creep in: for they love heat and humour that they find here. But against such adders a little beast fighteth that hight Saura, as it were a little ewt, and some men mean that it is a lizard; for when this beast is aware that this serpent is present, then he leapeth upon his face that sleepeth, and scratcheth with his feet to wake him, and to warn him of the serpent. And when this little beast waxeth old, his eyen wax blind, and then he goeth into an hole of a wall against the east, and openeth his eyen afterward when the sun is risen, and then his eyen heat and take light.

This slaying adder and venomous hath wit to love and affection, and loveth his mate as it were by love of wedlock, and liveth not well without company. Therefore if the one is slain, the other pursueth him that slew that other with so busy wreak and vengeance, that passeth weening. And knoweth the slayer, and reseth on him, be he in never so great company of men and of people, and busieth to slay him, and passeth all difficulties and spaces of ways, and with wreak of the said death of his mate. And is not let, ne put off, but it be by swift flight, or by waters or rivers. Marcianus saith that the asp grieveth not men of Africa or Moors; for they take their children that they have suspect, and put them to these adders: and if the children be of their kind, this adder grieveth them not, and if they be of other kind, anon they die by venom of the adder.

An oxherd hight Bubulcus, and is ordained by office to keep oxen: He feedeth and nourisheth oxen, and bringeth them to leas and home again: and bindeth their feet with a langhaldes and spanells and nigheth and cloggeth them while they be in pasture and leas, and yoketh and maketh them draw at the plough: and pricketh the slow with a goad, and maketh them draw even. And pleaseth them with whistling and with song, to make them bear the yoke with the better will for liking of melody of the voice. And this herd driveth and ruleth them to draw even, and teacheth them to make even furrows: and compelleth them not only to ear, but also to tread and to thresh. And they lead them about upon corn to break the straw in threshing and treading the flour. And when the travail is done, then they unyoke them and bring them to the stall: and tie them to the stall, and feed them thereat.

The cockatrice hight Basiliscus in Greek, and Regulus in Latin; and hath that name Regulus of a little king, for he is king of serpents, and they be afraid, and flee when they see him. For he slayeth them with his smell and with his breath: and slayeth also anything that hath life with breath and with sight. In his sight no fowl nor bird passeth harmless, and though he be far from the fowl, yet it is burned and devoured by his mouth. But he is overcome of the weasel; and men bring the weasel to the cockatrice's den, where he lurketh and is hid. For the father and maker of everything left nothing without remedy. Among the Hisperies and Ethiopians is a well, that many men trow is the head of Nile, and there beside is a wild beast that hight Catoblefas, and hath a little body, and nice in all members, and a great head hanging always toward the earth, and else it were great noying to mankind. For all that see his eyen, should die anon, and the same kind hath the cockatrice, and the serpent that is bred in the province of Sirena; and hath a body in length and in breadth as the cockatrice, and a tail of twelve inches long, and hath a speck in his head as a precious stone, and feareth away all serpents with hissing. And he presseth not his body with much bowing, but his course of way is forthright, and goeth in mean. He drieth and burneth leaves and herbs, not only with touch but also by hissing and blast he rotteth and corrupteth all things about him. And he is of so great venom and perilous, that he slayeth and wasteth him that nigheth him by the length of a spear, without tarrying; and yet the weasel taketh and overcometh him, for the biting of the weasel is death to the cockatrice. And nevertheless the biting of the cockatrice is death to the weasel. And that is sooth, but if the weasel eat rue before. And though the cockatrice be venomous without remedy, while he is alive, yet he loseth all the malice when he is burnt to ashes. His ashes be accounted good and profitable in working of Alchemy, and namely in turning and changing of metals.

Nothing is more busy and wittier than a hound, for he hath more wit than other beasts. And hounds know their own names, and love their masters, and defend the houses of their masters, and put themselves wilfully in peril of death for their masters, and run to take prey for their masters, and forsake not the dead bodies of their masters. We have known that hounds fought for their lords against thieves, and were sore wounded, and that they kept away beasts and fowls from their masters' bodies dead. And that a hound compelled the slayer of his master with barking and biting to acknowledge his trespass and guilt. Also we read that Garamantus the king came out of exile, and brought with him two hundred hounds, and fought against his enemies with wondrous hardiness.

Other hounds flee and avoid the wood hound as pestilence and venom: and he is always exiled as it were an outlaw, and goeth alone wagging and rolling as a drunken beast, and runneth yawning, and his tongue hangeth out, and his mouth drivelleth and foameth, and his eyes be overturned and reared, and his ears lie backward, and his tail is wrinkled by the legs and thighs; and though his eyes be open, yet he stumbleth and spurneth against every thing. And barketh at his own shadow.... Pliny saith that under the hound's tongue lieth a worm that maketh the hound wood, and if this worm is taken out of the tongue, then the evil ceaseth.... Also an hound is wrathful and malicious, so that for to awreak himself, he biteth oft the stone that is thrown to him: and biteth the stone with great woodness, that he breaketh his own teeth, and grieveth not the stone, but his own teeth full sore. Also he is guileful and deceivable, and so oft he fickleth and fawneth with his tail on men that pass by the way, as though he were a friend, and biteth them sore if they take none heed backward. And the hound hateth stones and rods, and is bold and hardy among them that he knoweth, and busieth to bite and to fear all other, and is not bold when he passeth among strangers. Also the hound is envious, and gathereth herbs privily, and is right sorry if any man know the virtue of those herbs, as is also evil apaid if any strange hounds and unknown come into the place where he dwelleth; and dreadeth lest he should fare the worse for the other hound's presence, and fighteth with him therefore. Also he is covetous and scarce, and busy to lay up and to hide the relief that he leaveth. And therefore he commoneth not, nor giveth flesh and marrow-bones that he may not devour to other hounds: but layeth them up busily, and hideth them until he hungereth again.... And at the last the hound is violently drawn out of the dunghill with a rope or with a whip bound about his neck, and is drowned in the river, or in some other water, and so he endeth his wretched life. And his skin is not taken off, nor his flesh is not eaten or buried, but left finally to flies, and to other divers worms.

In Pontus is a manner kind of beasts, that dwelleth now in land and now in water, and maketh houses and dens arrayed with wonder craft in the brinks of rivers and of waters. For these beasts live together in flocks, and love beasts of the same kind, and come together and cut rods and sticks with their teeth, and bring them home to their dens in a wonder wise, for they lay one of them upright on the ground, instead of a sled or of a dray, with his legs and feet reared upward, and lay and load the sticks and wood between his legs and thighs, and draw him home to their dens, and unlade and discharge him there, and make their dwelling places right strong by great subtlety of craft. In their houses be two chambers or three distinguished, as it were three cellars, and they dwell in the over place when the water ariseth, and in the nether when the water is away, and each of them hath a certain hole properly made in the cellar, by the which hole he putteth out his tail in the water, for the tail is of fishy kind, it may not without water be long kept without corruption.

If the crocodile findeth a man by the brim of the water, or by the cliff, he slayeth him if he may, and then he weepeth upon him, and swalloweth him at the last.

The Dragon is most greatest of all serpents, and oft he is drawn out of his den, and riseth up into the air, and the air is moved by him, and also the sea swelleth against his venom, and he hath a crest with a little mouth, and draweth breath at small pipes and straight, and reareth his tongue, and hath teeth like a saw, and hath strength, and not only in teeth, but also in his tail, and grieveth both with biting and with stinging, and hath not so much venom as other serpents: for to the end to slay anything, to him venom is not needful, for whom he findeth he slayeth, and the elephant is not secure of him, for all his greatness of body. Oft four or five of them fasten their tails together, and rear up their heads, and sail over sea and over rivers to get good meat. Between elephants and dragons is everlasting fighting, for the dragon with his tail bindeth and spanneth the elephant, and the elephant with his foot and with his nose throweth down the dragon, and the dragon bindeth and spanneth the elephant's legs, and maketh him fall, but the dragon buyeth it full sore: for while he slayeth the elephant, the elephant falleth upon him and slayeth him. Also the elephant seeing the dragon upon a tree, busieth him to break the tree to smite the dragon, and the dragon leapeth upon the elephant, and busieth him to bite him between the nostrils, and assaileth the elephant's eyen, and maketh him blind sometime, and leapeth upon him sometime behind, and biteth him and sucketh his blood. And at the last after long fighting the elephant waxeth feeble for great blindness, in so much that he falleth upon the dragon, and slayeth in his dying the dragon that him slayeth. The cause why the dragon desireth his blood, is coldness of the elephant's blood, by the which the dragon desireth to cool himself. Jerome saith, that the dragon is a full thirsty beast, insomuch that unneth he may have water enough to quench his great thirst; and openeth his mouth therefore against the wind, to quench the burning of his thirst in that wise. Therefore when he seeth ships sail in the sea in great wind, he flieth against the sail to take their cold wind, and overthroweth the ship sometimes for greatness of body, and strong rese against the sail. And when the shipmen see the dragon come nigh, and know his coming by the water that swelleth ayenge him, they strike the sail anon, and scape in that wise.

Horses be joyful in fields, and smell battles, and be comforted with noise of trumpets to battle and to fighting; and be excited to run with noise that they know, and be sorry when they be overcome, and glad when they have the mastery. And so feeleth and knoweth their enemies in battle so far forth that they a-rese on their enemies with biting and smiting, and also some know their own lords, and forget mildness, if their lords be overcome: and some horses suffer no man to ride on their backs, but only their own lords. And many horses weep when their lords be dead. And it is said that horses weep for sorrow, right as a man doth, and so the kind of horse and of man is medlied. Also oft men that shall fight take evidence and divine and guess what shall befall, by sorrow or by the joy that the horse maketh. Old men mean that in gentle horse, noble men take heed of four things, of shape, and of fairness, of wilfulness, and of colour.

In his forehead when he is foaled is found Iconemor, a black skin of the quantity of a sedge, that hight also Amor's Veneficium; and the mother licketh it off with her tongue, and taketh it away and hideth it or eateth it. For women that be witches use that skin in their sayings, when they will excite a man to love.... The colt is not littered with straw, nor curried with an horse comb, nor arrayed with trapping and gay harness, nor smitten with spurs, nor saddled with saddle, nor tamed with bridle, but he followeth his mother freely, and eateth grass, and his feet be not pierced with nails, but he is suffered to run hither and thither freely: but at the last he is set to work and to travail, and is held and tied and led with halters and reins, and taken from his mother, and may not suck his dam's teats; but he is taught in many manner wise to go easily and soft. And he is set to carts, chariots, and cars, and to travel and bearing of horsemen in chivalry: and so the silly horse colt is foaled to divers hap of fortune. Isidore saith, that horses were sometime hallowed in divers usage of the gods.

Among beasts the elephant is most of virtue, so that unneth among men is so great readiness found. For in the new moon they come together in great companies, and bathe and wash them in a river, and lowte each to other, and turn so again to their own places, and they make the young go tofore in the turning again; and keep them busily and teach them to do in the same wise: and when they be sick, they gather good herbs, and ere they use the herbs they heave up the head, and look up toward heaven, and pray for help of God in a certain religion. And they be good of wit, and learn well: and are easy to teach, insomuch that they be taught to know the king and to worship him, and busy to do him reverence and to bend the knees in worship of him. If elephants see a man coming against them that is out of the way in the wilderness, for they would not affray him, they will draw themselves somewhat out of the way, and then they stint, and pass little and little tofore him, and teach him the way. And if a dragon come against him, they fight with the dragon and defend the man, and put them forth to defend the man strongly and mightily: and do so namely when they have young foals, for they dread that the man seeketh their foals. And therefore they purpose first to deliver them of the man, that they may more securely feed their children and keep them the more warily.... Elephants be best in chivalry when they be tame: for they bear towers of tree, and throw down sheltrons, and overturn men of arms, and that is wonderful; for they dread not men of arms ranged in battle, and dread and flee the voice of the least sound of a swine. When they be taken, they be made tame and mild with barley: and a cave or a ditch is made under the earth, as it were a pitfall in the elephant's way, and unawares he falleth therein. And then one of the hunters cometh to him and beateth and smiteth him, and pricketh him full sore. And then another hunter cometh and smiteth the first hunter, and doth him away, and defendeth the elephant, and giveth him barley to eat, and when he hath eaten thrice or four times, then he loveth him that defended him, and is afterward mild and obedient to him. I have read in Physiologus' book that the elephant is a beast that passeth all other four-footed beasts in quantity, in wit, and in mind. For among other doings elephants lie never down in sleeping; but when they be weary they lean to a tree and so rest somewhat. And men lie in wait to espy their resting places privily, for to cut the tree in the other side: and the elephant cometh and is not aware of the fraud, and leaneth to the tree and breaketh it with the weight of his body, and falleth down with the breaking, and lieth there. And when he seeth he may not help himself in falling he crieth and roareth in a wonder manner: and by his noise and crying come suddenly many young elephants, and rear up the old little and little with all their strength and might: and while they arear him with wonder affection and love, they bend themselves with all their might and strength. ... Also there is another thing said that is full wonderful: among the Ethiopians in some countries elephants be hunted in this wise: there go in the desert two maidens all naked and bare, with open hair of the head: and one of them beareth a vessel, and the other a sword. And these maidens begin to sing alone: and the beast hath liking when he heareth their song, and cometh to them, and licketh their teats, and falleth asleep anon for liking of the song, and then the one maid sticketh him in the throat or in the side with a sword, and the other taketh his blood in a vessel, and with that blood the people of the same country dye cloth, and done colour it therewith.

Satyrs be somewhat like men, and have crooked nose and horns in the forehead, and like to goats in their feet. Saint Anthony saw such a one in the wilderness, as it is said, and he asked what he was, and he answered Anthony, and said: "I am deadly, and one of them that dwelleth in the wilderness." These wonderful beasts be divers: for some of them be called Cyno[ce]phali, for they have heads as hounds, and seem by the working, beasts rather than men, and some be called Cyclops, and have that name, for one of them hath but one eye, and that in the middle of the forehead, and some be all headless and noseless, and their eyen be in the shoulders, and some have plain faces without nostrils, and the nether lips of them stretch so, that they hele therewith their faces when they be in the heat of the sun: and some of them have closed mouths, in their breasts only one hole, and breathe and suck as it were with pipes and veins, and these be accounted tongueless, and use signs and becks instead of speaking. Also in Scythia be some with so great and large ears, that they spread their ears and cover all their bodies with them, and these be called Panchios.... And other be in Ethiopia, and each of them have only one foot so great and so large, that they beshadow themselves with the foot when they lie gaping on the ground in strong heat of the sun; and yet they be so swift, that they be likened to hounds in swiftness of running, and therefore among the Greeks they be called Cynopodes. Also some have the soles of their feet turned backward behind the legs, and in each foot eight toes, and such go about and stare in the desert of Lybia. The griffin is a beast with wings, and is four footed: and breedeth in the mountains Hyperborean, and is like to the lion in all the parts of the body, and to the eagle only in the head and wings. And griffins keep the mountains in which be gems and precious stones, and suffer them not to be taken from thence.

The hyena is a cruel beast like to the wolf in devouring and gluttony, and reseth on dead men, and taketh their carcase out of the earth, and devoureth them. It is his kind to change sex, for he is now found male, and now female, and is therefore an unclean beast, and cometh to hoveys by night, and feigneth man's voice as he may, for men should trow that it is a man. Pliny saith: It is said he is one year male and another female. And she bringeth forth her brood without male, as the common people trow. But Aristotle denieth that. And hath the neck of the adder, and the ridge of an elephant, and may not bend but if he bear all the body about. And herds tell that among stables, he feigneth speech of mankind, and calleth some man by his own name, and rendeth him when he hath him without. And he feigneth oft the name of some man, for to make hounds run out, that he may take and eat them.... And his shadow maketh hounds leave barking and be still, if he come near them. And if this beast hyena goeth thrice about any beast, that beast shall stint within his steps. Pliny saith that the hyena hateth the panther. And it is said that if both their skins be hanged together, the hair of the panther's skin shall fall away. This beast hyena fleeth the hunter, and draweth toward the right side, to occupy the trace of the man that goeth before: and if he cometh not after, he telleth that he goeth out of his wit, or else the man falleth down off his horse. And if he turn against the hyena, the beast is soon taken, as magicians tell. And also witches use the heart of this beast and the liver, in many witchcrafts.

Some lions be short with crisp hair and mane, and these lions fight not; and some lions have simple hair of mane, and those lions have sharp and fierce hearts, and by their foreheads and tails their virtue is known in the beast, and their stedfastness in the head: and when they be beset with hunters, then they behold the earth, for to dread the less the hunters and their gins, that them have beset about: and he dreadeth noise and rushing of wheels, but he dreadeth fire much more. And when they sleep their eyes wake: and when they go forth or about, they hele and hide their fores and steps, for hunters should not find them.... It is the kind of lions, not to be wroth with man, but if they be grieved or hurt. Also their mercy is known by many and oft examples: for they spare them that lie on the ground, and suffer them to pass homeward that were prisoners and come out of thraldom, and eat not a man or slay him but in great hunger. Pliny saith that the lion is in most gentleness and nobility, when his neck and shoulders be heled with hair and main. And he that is gendered of the pard, lacketh that nobility. The lion knoweth by smell, if the pard gendereth with the lioness, and reseth against the lioness that breaketh spousehood, and punisheth her full sore, but if she wash her in a river, and then it is not known. The lion liveth most long, and that is known by working and wasting of his teeth: and when in age he reseth on a man: for his virtue and might faileth to pursue great beasts and wild. And then he besiegeth cities to ransom and to take men: but when the lions be taken, then they be hanged, for other lions should dread such manner pain. The old lion reseth woodly on men, and only grunteth on women, and reseth seldom on children, but in great hunger.... In peril the lion is most gentle and noble, for when he is pursued with hounds and with hunters, the lion lurketh not nor hideth himself, but sitteth in fields where he may be seen, and arrayeth himself to defence. And runneth out of wood and covert with swift running and course, as though he would account vile shame to lurk and to hide himself. And he hideth himself not for dread that he hath, but he dreadeth himself sometime, only for he would not be dreaded. And when he pursueth man or beast in lands, then he leapeth when he reseth on him. When he is wounded, he taketh wonderly heed, and knoweth them that him first smiteth, and reseth on the smiter, though he be never in so great multitude: and if a man shoot at him, the lion chaseth him and throweth him down, and woundeth him not, nor hurteth him.... He hideth himself in high mountains, and espieth from thence his prey. And when he seeth his prey he roareth full loud, and at the voice of him other beasts dread and stint suddenly: and he maketh a circle all about them with his tail, and all the beasts dread to pass out over the line of the circle, and the beasts stand astonied and afraid, as it were abiding the hest and commandment of their king.... And he is ashamed to eat alone the prey that he taketh; therefore of his grace of free heart, he leaveth some of his prey to other beasts that follow him afar.... And the lion is hunted in this wise: One double cave is made one fast by that other, and in the second cave is set a whiche, that closeth full soon when it is touched: and in the first den and cave is a lamb set, and the lion leapeth therein, when he is an hungered, for to take the lamb. And when he seeth that he may not break out of the den, he is ashamed that he is beguiled, and would enter in to the second den to lurk there, and falleth smell, if the pard gendereth with the lioness, and reseth against the lioness that breaketh spousehood, and punisheth her full sore, but if she wash her in a river, and then it is not known. The lion liveth most long, and that is known by working and wasting of his teeth: and when in age he reseth on a man: for his virtue and might faileth to pursue great beasts and wild. And then he besiegeth cities to ransom and to take men: but when the lions be taken, then they be hanged, for other lions should dread such manner pain. The old lion reseth woodly on men, and only grunteth on women, and reseth seldom on children, but in great hunger.... In peril the lion is most gentle and noble, for when he is pursued with hounds and with hunters, the lion lurketh not nor hideth himself, but sitteth in fields where he may be seen, and arrayeth himself to defence. And runneth out of wood and covert with swift running and course, as though he would account vile shame to lurk and to hide himself. And he hideth himself not for dread that he hath, but he dreadeth himself sometime, only for he would not be dreaded. And when he pursueth man or beast in lands, then he leapeth when he reseth on him. When he is wounded, he taketh wonderly heed, and knoweth them that him first smiteth, and reseth on the smiter, though he be never in so great multitude: and if a man shoot at him, the lion chaseth him and throweth him down, and woundeth him not, nor hurteth him.... He hideth himself in high mountains, and espieth from thence his prey. And when he seeth his prey he roareth full loud, and at the voice of him other beasts dread and stint suddenly: and he maketh a circle all about them with his tail, and all the beasts dread to pass out over the line of the circle, and the beasts stand astonied and afraid, as it were abiding the hest and commandment of their king.... And he is ashamed to eat alone the prey that he taketh; therefore of his grace of free heart, he leaveth some of his prey to other beasts that follow him afar.... And the lion is hunted in this wise: One double cave is made one fast by that other, and in the second cave is set a whiche, that closeth full soon when it is touched: and in the first den and cave is a lamb set, and the lion leapeth therein, when he is an hungered, for to take the lamb. And when he seeth that he may not break out of the den, he is ashamed that he is beguiled, and would enter in to the second den to lurk there, and falleth into it, and it closeth anon as he is in, and letteth him not pass out thereof, but keepeth him fast therein, until he be taken out and bound with chains till he be tame.... The lion is cruel and wood when he is wroth, and biteth and grieveth himself for indignation, and gnasheth with his teeth, and namely when he hungreth, and spieth and lieth in wait, to take beasts which pass by the way. He hideth himself in privy caves, and reseth on beasts unawares, and slayeth them with his teeth and claws, and breaketh all their members, and eateth them piecemeal: and if he see any come against him to take away his prey, then he beclippeth the prey, and grunteth and smiteth the earth with his tail, and if he nigheth him he leapeth on him, and overcometh him, and turneth to the prey. First he drinketh and licketh the blood of the beast that he slayeth, and rendeth and haleth the other-deal limb- meal, and devoureth and swalloweth it.

The leopard is a beast most cruel, and is gendered in spouse-breach of a pard and of a lioness, and pursueth his prey startling and leaping and not running, and if he taketh not his prey in the third leap, or in the fourth, then he stinteth for in-dignation, and goeth backward as though he were overcome. And he is less in body than the lion, and therefore he dreadeth the lion, and maketh a cave under earth with double entering, one by which he goeth in, and the other by which he goeth out. And that cave is full wide and large in either entering, and more narrow and straight in the middle. And so when the lion cometh, he fleeth and falleth suddenly into the cave, and the lion pursueth him with a great rese, and entereth also into the cave, and weeneth there to have the mastery over the leopard, but for greatness of his body he may not pass freely by the middle of the den which is full straight, and when the leopard knoweth that the lion is so let and holden in the straight place, he goeth out of the den forward, and cometh again into the den in the other side be-hind the lion, and reseth on him behindforth with biting and with claws, and so the leopard hath often in that wise the mastery of the lion by craft and not by strength, so the less beast hath oft the mastery of the strong beast by deceit and guile in the den, and dare not rese on him openly in the field, as Homer saith in the book of the battles and wiles of beasts.

Churls speak of him [the wolf] and say that a man loseth his voice, if the wolf seeth him first. Therefore to a man that is suddenly still, and leaveth to speak it is said, "Lupus est in fabula," "The wolf is in the tale." And certainly if he know that he is seen first, he loseth his boldness, hardihood, and fierceness. The wolf is an evil beast, when he eateth, and resteth much when he hath no hunger: he is full hardy, and loveth well to play with a child, if he may take him; and slayeth him afterward, and eateth him at the last. It is said, that if the wolf be stoned, he taketh heed of him that threw the first stone, and if that stone grieveth him he will slay him: and if it grieveth him not, and he may take him that throweth that stone, he doth him not much harm, but some harm he doth him as it were in wrath, and leaveth him at last.... The wolf may not bend his neck backward in no month of the year but in May alone, when it thundereth. And when he goeth by night to a fold for to take his prey, he goeth against the wind for hounds should not smell him. And if it happeth in any wise that his foot maketh noise, treading upon anything, then he chasteneth that foot with hard biting.... I have read in a book that a string made of a wolf's gut, put among harp strings made of the guts of sheep, destroyeth and corrupteth them, as the eagle's feathers put among culvours', pulleth and gnaweth them, if they be there left together long in one place.

He [the cat] is a full lecherous beast in youth, swift, pliant, and merry, and leapeth and reseth on everything that is to fore him: and is led by a straw, and playeth therewith: and is a right heavy beast in age and full sleepy, and lieth slyly in wait for mice: and is aware where they be more by smell than by sight, and hunteth and reseth on them in privy places: and when he taketh a mouse, he playeth therewith, and eateth him after the play. In time of love is hard fighting for wives, and one scratcheth and rendeth the other grievously with biting and with claws. And he maketh a ruthful noise and ghastful, when one proffereth to fight with another: and unneth is hurt when he is thrown down off an high place. And when he hath a fair skin, he is as it were proud thereof, and goeth fast about: and when his skin is burnt, then he bideth at home; and is oft for his fair skin taken of the skinner, and slain and flayed.

Physiologus speaketh of the Panther and saith that he hateth the dragon, and the dragon fleeth him: and when he hath eat enough at full, he hideth him in his den, and sleepeth continually nigh three days, and riseth after three days and cri- eth, and out of his mouth cometh right good air and savour, and is passing measure sweet: and for the sweetness all beasts follow him. And only the dragon is a- feared when he heareth his voice, and fleeth into a den, and may not suffer the smell thereof; and faileth in himself, and looseth his comfort. For he weeneth that his smell is very venom.

All four-footed beasts have liking to behold the divers colours of the panther and tiger, but they are a-feared of the horribleness of their heads, and therefore they hide their heads, and toll the beasts to them with fairness of that other-deal of the body, and take them when they come so tolled, and eat them.

The mermaid is a sea beast wonderly shapen, and draweth shipmen to peril by sweetness of song. The Gloss on Is. xiii. saith that sirens are serpents with crests. And some men say, that they are fishes of the sea in likeness of women. Some men feign that there are three Sirens some-deal maidens, and some-deal fowls with claws and wings, and one of them singeth with voice, and another with a pipe, and the third with an harp, and they please so shipmen, with likeness of song, that they draw them to peril and to shipbreach, but the sooth is, that they were strong hores, that drew men that passed by them to poverty and to mischief. And Physiologus saith it is a beast of the sea, wonderly shapen as a maid from the navel upward and a fish from the navel downward, and this wonderful beast is glad and merry in tempest, and sad and heavy in fair weather. With sweetness of song this beast maketh shipmen to sleep, and when she seeth that they are asleep, she goeth into the ship, and ravisheth which she may take with her, and bringeth him into a dry place, and maketh him first lie by her, and if he will not or may not, then she slayeth him and eateth his flesh. Of such wonderful beasts it is written in the great Alexander's story.

The tiger is the swiftest beast in flight, as it were an arrow, for the Persees call an arrow Tigris, and is a beast distinguished with divers specks, and is wonderly strong and swift. And Pliny saith that they be beasts of dreadful swiftness, and that is namely known when he is taken, for the whelp is all glimy and sinewy; and the hunter lieth in await, and taketh away the whelps, and fleeth soon away on the most swift horse that he may have. And when the wild beast cometh and findeth the den void, and the whelps away, then he reseth headlong, and taketh the fore of him that beareth the whelps away, and followeth him by smell, and when the hunter heareth the grutching of that beast that runneth after him, he throweth down one of the whelps; and the mother taketh the whelp in her mouth, and beareth him into her den and layeth him therein, and runneth again after the hunter. But in the meantime the hunter taketh a ship, and hath with him the other whelps, and scapeth in that wise; and so she is beguiled and her fierceness standeth in no stead, and the male taketh no wood rese after. For the male recketh not of the whelps, and he that will bear away the whelps, leaveth in the way great mirrors, and the mother followeth and findeth the mirrors in the way, and looketh on them and seeth her own shadow and image therein, and weeneth that she seeth her children therein, and is long occupied therefore to deliver her children out of the glass, and so the hunter hath time and space for to scape, and so she is beguiled with her own shadow, and she followeth no farther after the hunter to deliver her children.

Avicenna saith that the bear bringeth forth a piece of flesh imperfect and evil shapen, and the mother licketh the lump, and shapeth the members with licking.... For the whelp is a piece of flesh little more than a mouse, having neither eyes nor ears, and having claws some-deal bourgeoning, and so this lump she licketh, and shapeth a whelp with licking.... And it is wonder to tell a thing, that Theophrastus saith and telleth that bear's flesh sodden that time (of their sleeping) vanisheth if it be laid up, and is no token of meat found in the almery, but a little quantity of humour.... When he is taken he is made blind with a bright basin, and bound with chains, and compelled to play, and tamed with beating; and is an unsteadfast beast, and unstable and uneasy, and goeth therefore all day about the stake, to the which he is strongly tied. He licketh and sucketh his own feet, and hath liking in the juice thereof. He can wonderly sty upon trees unto the highest tops of them, and oft bees gather honey in hollow trees, and the bear findeth honey by smell, and goeth up to the place that the honey is in, and maketh a way into the tree with his claws, and draweth out the honey and eateth it, and cometh oft by custom unto such a place when he is an-hungered. And the hunter taketh heed thereof, and pitcheth full sharp hooks and stakes about the foot of the tree, and hangeth craftily a right heavy hammer or a wedge tofore the open way to the honey. And then the bear cometh and is an- hungered, and the log that hangeth there on high letteth him: and he putteth away the wedge despiteously, but after the removing the wedge falleth again and hitteth him on the ear. And he hath indignation thereof, and putteth away the wedge despiteously and right fiercely, and then the wedge falleth and smiteth him harder than it did before, and he striveth so long with the wedge, until his feeble head doth fail by oft smiting of the wedge, and then he falleth down upon the pricks and stakes, and slayeth himself in that wise. Theophrastus telleth this manner hunting of bears, and learned it of the hunters in the country of Germany.

A fox is called Vulpes, and hath that name as it were wallowing feet aside, and goeth never forthright, but always aslant and with fraud. And is a false beast and deceiving, for when him lacketh meat, he feigneth himself dead, and then fowls come to him, as it were to a carrion, and anon he catcheth one and devoureth it. The fox halteth always, for the right legs are shorter than the left legs. His skin is right hairy rough and hot, his tail is great and rough; and when an hound weeneth to take him by the tail, he taketh his mouth full of hair and stoppeth it. The fox doth fight with the brock for dens, and defileth the brock's den, and hath so the mastery over him with fraud and deceit, and not by strength.... The fox feigneth himself tame in time of need, but by night he waiteth his time and doeth shrewd deeds. And though he be right guileful in himself and malicious, yet he is good and profitable in use of medicine.

# THE SOURCES OF THE BOOK

ADAMANTIUS (fl. 258). Origen it quoted under this name. His commentaries on the Old Testament are the works quoted from.

AEGIDIUS CORBOLIENSIS, of Corbeil (d. 1220). A doctor at Montpellier, and Canon of Paris.

ALANUS DE INSULIS, OR DE RYSSEL (d. 1202). A monk of Canterbury, most probably an Englishman. His principal work is a poem in 9 books, called ANTI-CLAUDIANUS, largely quoted by all Middle Age writers. An account of it is given in the notes on the Secreta Secretorum (E.E.T.S.). He also wrote DE PLANCTU NATURAE, PARABOLAE, etc.

ALBERTUS MAGNUS (1193-1280). A famous doctor in the University of Paris and a Dominican Theologian. The works quoted are commentaries on the Natural Histories of Aristotle. They have often been printed. He was teacher of Thomas Aquinas and a contemporary of our author.

ALBUMAZAR (d. 886). An Arab astronomer.

ALCUIN (735-804). An English theologian: the work quoted is his "De Septem Artibus."

ALEXANDER NECKHAM, OR NEQUAM (1157-1217). His principal work is "De Naturis Rerum," a book little known on the Continent. Its use by Bartholomew is thus another proof of his English birth.

ALFARAGUS (9th cent.). An Arab astronomer, whose work is notable as being the chief source of the celebrated astronomical treatise, "The Sphere," of Johannes Sacrobosco (John of Halifax), a contemporary Englishman. It was the popular text-book for over three centuries, and was as well known as Euclid.

ALFREDUS ANGLICUS (fl. 1200). A physician and translator of Aristotle. See JACOB'S AESOP for a discussion on his works.

AL GHAZEL (1061-1137). A sceptic opponent of Averroes.

AMBROSE (d. 397). The Hexameron is the work used.

ANSELM (1033-1109). Theologian, Archbishop of Canterbury. The inventor of Scholasticism.

ARCHELAUS. A Greek geographer.

ARISTOTLE (384-322 B.C.). I would refer the reader to BRÈCHILLET JOUR-DAIN on the EARLY TRANSLATIONS OF ARISTOTLE, where he will find a mine of information on the works of this writer used in the Middle Age.

AUGUSTINE (d. 430).

AURORA, THE. A metrical version of the Bible by PETRUS DE RIGA, Canon of Rheims (d. 1209).

AVERROES (d. 1217). Moorish commentator on Aristotle.

AVICEBRON (d. 1070), OR IBN GEBIROL. A Spanish Jew. Author of the FONTIS VITA. A work translated by Gundisalvi, of the greatest influence on the Metaphysic of the Middle Age. See MUNCK, MÉLANGES.

AVICENNA (980-1036). An Arab physician, and commentator on Aristotle.

AYMON, OR HAYMON (d. 1244). An English Franciscan, afterwards General of the Order, who revised the breviary and rubrics.

BASIL (329-379). In HEXAMERON.

BEDE (673-735). The work by which he was best known in the thirteenth century was not his History but the works on the *Calendar*, etc.

BELETH, JOHN (before 1165). A French writer on ecclesiastical matters.

BERNARD (1091-1153).

BESTIARIUM. A collection of early myths on animals; of Eastern origin. There are many different forms of this work. All are founded on Physiologus.

BOETHIUS (470-526). His treatise on arithmetic is the work quoted here. His "Consolation" was almost unknown in the early Middle Age, his popularity resting on his translations of Aristotle and his treatises on Music and Arithmetic, the latter being a very important work in the history of the science.

CALLISTHENES, PSEUDO-. Author of the HISTORIA ALEXANDRI MAGNI DE PRELIIS. See BUDGE'S Syriac Version of this work.

CASSIODORUS (480-575). DE SEPTEM DISCIPLINIS. One of the favourite Middle Age Text-Books.

CATO (233-151 B.C.). On AGRICULTURE.

CHALCIDIUS (3rd cent.). A commentator on the TIMAEUS of Plato. Only a part of this is preserved.

CICERO (107-44 B.C.). In SOMN. SCIPIONIS.

CONSTANTINUS AFER (d. 1087). A Benedictine monk of Monte Cassino, and most probably the introducer of Arab medicine into Italy. He wrote the VIATI-CUM and the PANTEGNA (20 books). He introduced Arab medicine into Europe through the School of Salerno, translating many Arab authors.

CYPRIAN (d. 285). A Syriac astrologer, afterwards Bishop of Antioch, and Martyr in the Diocletian persecution.

DAMASCENE (11th cent.). Quoted by Constantinus Afer. A physician.

DAMASCENE, JOHN (end of 12th cent.). An Arab physician.

DAMASCIUS (circ. 533). A Syrian commentator on Aristotle, who took refuge in Persia. Author of a work on wonders quoted by Photius.

DIOSCORIDES (d. 47 B.C.).

DIONYSIUS AREOPAGITUS, PSEUDO- (circ. 400). DE CELESTI HIERAR-CHIA, DE DIVINIS NOMINIBUS.

DONATUS (333). A Grammarian.

EUFICIUS (circ. 600). A disciple of Gregory.

FULGENTIUS (circ. 550). A grammarian.

GALEN (131-210).

GILBERTUS (circ. 1250). A celebrated English physician in France; wrote COMPENDIUM MEDICINAE.

GREGORY (circ. 590). On Job.

HALY (circ. 1000). A Jewish physician. Wrote a PANTEGNI or COMPLE-MENTUM MEDICINAE. The first medical work translated by Constantius Afer.

HERMES. In ALCHEMIA (not now extant).

HIPPOCRATES (460-351 B.C.).

HUGUTION PIZANUS (d. 1210). A jurisconsult and writer on Grammar.

HYGINUS, PSEUDO- (6th cent.). Writer on Astronomy.

INNOCENT III. (d. 1216). Wrote "De Contemptu Mundi," etc.

ISAAC (circ. 660). An Arab physician, who translated many Greek authors into Arabic.

ISIDORE (d. 636). Bishop of Seville. He wrote a work on Etymology in 20 books, one of the most popular works of the Middle Age.

JACOBUS DE VITRIACO (d. 1240). A Crusading Bishop, afterwards Cardinal legate. Wrote an EXEMPLAR, and 3 books of Eastern and Western History.

JEROME (340-420).

JOSEPH BEN GORION (900). Abridgment of Jewish History containing many legends.

JOSEPHUS (37-95). Jewish historian.

JORATH. DE ANIMALIBUS. A Syriac writer (?).

LAPIDARIUM. See MARBODIUS DE GEMMIS. There are many treatises under this name.

LEO IX. (1054). See Migne, Patrologia.

LUCAN (d. 65). One of the most popular Latin poets of the Middle Age.

MACER FLORIDUS (6th cent.). On THE VIRTUES OF HERBS.

MACROBIUS (circ. 409). His commentary on the dream of Scipio was a favourite work in Medieval times.

MARTIANUS CAPELLA (circ. 400). Wrote a poem, THE MARRIAGE OF MERCURY AND PHILOLOGIA, treating of THE SEVEN LIBERAL ARTS, which was the standard text-book from the 5th century for the schools.

MESSAHALA (circ. 1100).

METHODIUS, PSEUDO- (8th cent.). DE AGARINI.

MICHAEL SCOT (circ. 1235). At this time concerned in the translation of some Arabic works on Astronomy, and Aristotle's DE COELO and DE MUNDO DE ANIMA, and HISTORIA NATURALIS with commentaries.

MISALATH ASTROLOGUS (?).

PAPIAS (circ. 1053). Grammarian. [Milan, 1467, etc.]

PERSPECTIVA SCIENCIA. I cannot say whether this is Bacon's, Peckham's, or Albertus Magnus', but I believe it to be Peckham's, who was an Englishman, and afterwards Archbishop of Canterbury.

PETRUS COMESTOR (d. 1198). Named MAGISTER HISTORIARUM or Master of Histories, wrote an account of the world from the Creation, which, when translated into French, was called the "Mer des Histoires." A favourite Medieval book.

PHILARETUS (1100). A writer on Medicine.

PHYSIOLOGUS. A Syriac compilation of moralities on animal myths. It first appears in Western Europe as THEOBALDUS DE NATURIS XII. ANIMALIUM. Of Alexandrian origin, it dates from before the fourth century, and appears to have been altered at the will of each writer.

PLATEARIUS SALERNITANUS (circ. 1100) was Johannes, one of a family of physicians at Salerno. His work is called the PRACTICA. A book on the virtues of herbs. [Lugd., 1525, etc.]

PLATO (430-348 B.C.). The TIMAEUS is quoted, probably from Chalcidius.

PLINY (d. 79). Natural History. This and Isidore's work are the two chief sources of medieval knowledge of Nature.

PRISCIAN (circ. 525). Grammarian and physicist.

PTOLEMY (circ. 130). An Alexandrian astronomer, known through Arabic translations only at that time. [Ven., 1509, etc.]

RABANUS MAURUS (776-856) of Fulda, pupil of Alcuin. A Benedictine, afterwards Archbishop of Mayence, who wrote DE UNIVERSO MUNDO. [1468; Col., 1627, etc.]

RASIS (d. 935). An Arab physician, perhaps the greatest of the School. [Ven., 1548, etc.]

REMIGIUS (d. 908). A teacher of Grammar in the School of Paris. His grammar remained in use there four centuries. He wrote a gloss on Marcianus Capella.

RICARDUS DE ST. VICTOR (d. 1173). A Scottish theologian, Prior of St. Victor. A mystic of considerable acuteness. [Ven., 1506, etc.]

RICARDUS RUFUS (circ. 1225). A Cornishman who was a doctor in great renown, both at Oxford and Paris. He afterwards joined the Franciscans.

ROBERTUS LINCOLN., GROSTÊTE (d. 1253), the celebrated Bishop of Lincoln and patron of Bacon. Taught at Paris and at Oxford. Commentaries on Aristotle.

SALUSTIUS (d. 363?). DE DIIS ET MUNDO. A geographer.

SCHOLA SALERNITANA (circ. 1100). A treatise on the preservation of health in leonine verse for popular use, said to be addressed to Robert of England. It has been translated and commented on hundreds of times. The Middle Age very sensibly thought preservation from disease a branch of medicine equally important with the cure of it.

SECUNDUS. A writer on Medicine.

SOLINUS (circ. 100). Wrote an account of things in general-- POLYHISTO-RIA.

STEPHANUS (circ. 600). Commentary on Galen.

STRABUS (d. 847). A Benedictine, Abbot of Reichenau, near Constance. One of the authors of the Gloss.

SYMON CORNUBIENSIS (?).

VARRO, M. T. (116-26 B.C.). Most celebrated grammarian.

VIRGIL (70-19 B.C.).

WILLIAM CONCHES (d. 1150). Lectured at Paris, 1139, on Grammar, wrote DE NATURA.

ZENO (circ. 400), A writer on Medicine, and teacher at Alexandria.

This list of Authorities cited is that given at the end of the complete work of Bartholomew.

# BIBLIOGRAPHY

The first edition of this selection was published at London in 1893. The 1535 edition has 8 unpaged leaves (title, table, prologue, and Book I.), 338 numbered leaves, and printer's mark of Lucretia. The following errors in pagination are noted: 181 for 189, 197 for 187, 201 for 200, 203 for 201, 211 for 209.

The chief point of interest in the Bibliography is the question raised by Wynkyn de Worde's positive statement in his edition in his epilogue:

> And also of your charyte call to remembraunce
> The soule of William Caxton first prynter of this boke
> In latin tonge at Coleyn hymself to avaunce
> That every well disposyd man may theron loke
> And John Tate the yonger Joy mote he broke
> Which late hathe in Englond doo make this paper thynne
> That now in our Englyssh this boke is prynted Inne.

Mr. Gordon Duff is disposed to think that Caxton may have worked on the undated Cologne edition (H.C. *2498), which must in that case be put before 1476, finding a link between his Bruges type and the Cologne presses in a work printed at Louvain in 1475 which contains type of both descriptions.

Most of these editions are in the British Museum. The copy of the Berthelet edition there has an autograph of Shakespeare in it--one of the Ireland forgeries.

# GLOSSARY

Accord, **n.**, harmony
According, **part.**, punning, or in harmony
Adamant, **n.**, a diamond
Addercop, **n.**, a spider
Afeard, **part.**, affrighted
Afore, **prep.**, before
Almery, **n.**, a cupboard, a buttery
Anon, **adv.**, immediately
Apaid, **v.**, served, repaid
Apaired, **adj.**, injured, impaired
Areared, **adj.**, upright
Assay, **v.**, to try
Aught, **n.**, anything
Avisement, **n.**, forethought, counsel
Away with, **v.**, to suffer
Awreak, **v.**, revenge
Ayencoming, **n.**, returning
Ayenge, **prep.**, against
Ayenward, **adv.**, vice versa

Bate, **v.**, **hawking**, to flutter the wings as if preparing
for flight
Bays, **n.**, the fruit of the laurel

Because, *conj.*, in order that

Beclip, *v.*, embrace, enfold

Behind forth, *adv.*, from back to front

Behooteth, *v.*, advises, gives

Behove, *v.*, to be necessary

Bernacle, *n.*, a bridle

Beshine, *v.*, to illuminate

Bisse, *n.*, a second

Blemish, *v.*, shrink, blench

Blow, *v.*, to obtain lead, etc., from ores in a furnace

Boisterous, boystous, *adj.*, thick, strong, solid

Bourgeon, *v.*, to bud, burst forth

Bray, *v.*, to pound

Brock, *n.*, a badger

Buck, *v.*, to wash

Busily, *adv.*, carefully

But, *prep.*, except

Car, *n.*, means or instrument

Carfle, *v.*, to pound

Carrions, *n.*, corpses

Cast, *v.*, to intend

Chaffer, *n.*, trade

Chine, *n.*, chink, cleft

Clarity, *n.*, clearness

Clepe, *v.*, call

Cliff, *n.*, shore

Clue, *n.*, a clew or hank (of yarn)

Comfort, *v.*, to strengthen

Common, *v.*, to share one's food with others and ayenward

Conject, *v.*, conjecture

Coverture, *n.*, covering
Craftily, *adv.*, skilfully
Culvour, *n.*, pigeon
Curtel, *n,*, a kirtle, a short coat, a covering

Deadly, *adv.*, mortal
Deeming, *n.*, judgment, opinion
Default, *n.*, deficiency
Depart, *v.*, to separate, share out
Despiteously, *adv.*, contemptuously
Detty, *adj.*, generous
Disperple, *v.*, to scatter, destroy
Do, done, *v.*, to put, to don
Doomsman, *n.*, judge
Draust, *n.*, dross, impurity

Ear, *v.*, to reap
Else, *adv.*, otherwise
Enform, *v.*, to make
Even tofore, *adv.*, opposite to
Expert, *adv.*, tried

Fare, *v.*, to happen
Fear, *v. a.*, to frighten
Fell, *n.*, an undressed skin
Fen, *n.*, clay
Fine, *n.*, a boundary
Fleet, *v.*, to float, to swim; *cf.* "to flit"
Flux, *n.*, a flow, a catarrh
Fore, *n.*, trail, spoor; *cf.* "foor"
Frot, *v.*, to rub

Fumous, *adj.*, vaporous, cloudy

Fumosity, *n.*, vapour

Fundament, *n.*, foundation

Gentle, *adj.*, noble, high-minded

Gesses, *n.*, jesses, cords for fastening the legs of a hawk

Gete, *n.*, goats

Ghastful, *adj.*, frightful

Gin, *a.*, machine

Glad, *v. a.*, to please

Glimy, *adj.*, slimy

Gloss, *n.*, the comment on Scripture, compiled in the ninth century from the fathers

Glue, *n.*. any glutinous substance

Gnod, *v.*, to rub?

Grieve, *v.*, to hurt

Grutching, *n.*, growling

Gutter, *n.*, drop

Hale, *v.*, to drag

Hap, *n.*, chance

Hards, hirds, *n.*, tow

Haught, *part.*, hatched

Heckle, *v.*, to straighten out lint by a coarse comb

Hele, *v.*, to cover; *cf.*, heling

Hight, *v.*, is called

Hoar, *adj.*, feathered

Hop, *n.*, the seed case of the flaxplant

Horrible, *adj.*, unpleasant to hear

Housebond, *n.*, husband

Hovey, *part.*, hovel, cottage
Hoving, *part.*, staying

Infect, *adj.*, spotted, injured
Intendment, *n.*, understanding

Jape, *v.*, to cry out

Kele, *v.*, to cool
Kind, *n.*, nature
Kindly, *adj.*, natural; *adv.*, naturally

Langhaldes, *n.*, ropes connecting the fore and hind legs of a
horse or cow to stay it from jumping
Latten, *n.*, a kind of brass
Lea, *n.*, pasture land
Lesings, *n.*, untruths
Let, *v.*, to hinder
Lewd, *adj.*, ignorant
Liefer, *adv.*, rather
Likelihood, *n.*, resemblance
Limb, *n.*, an instrument; *cf.*, "limb of the law"
Limbmeal, *adv.*, limb by limb; *cf.*, "piecemeal"
List, *n.*, a limit, border
Lodesman, *n.*, pilot
Lowte, *v.*, to trumpet

Make, *n.*, a mate
Manner, *adj.*, manner of, kind of
Mawmet, *n.*, an idol or toy
Mean, *n.*, intermediary, means

Mean, *v.*, to assert, consider

Medley, *v.*, to mix

Meinie, *n.*, domestics, household

Merry, *adj.*, fortunate

Meselry, *n.*, leprosy Mess,n., portion

Messager, *n.*, messenger

Mete, *v.*, measure, apportion

Mews, *n.*, originally a place in which hawks were kept "mewed up"

Mildness, *n.*, generosity

Minish, *v.*, to narrow

Mirror, *n.*, seems to have been used only when the surface was curved, the word "shewer" being used for a plane mirror

Mistake, *v.*, to take wrongly


Namely, *adj.*, especially

Nathless, *con.*, nevertheless

Ne, *con.*, nor

Needly, *adj.*, necessarily

Nerve, *n.*, sinew

Nesh, *adj.*, soft

Nether, *adj.*, lower

Nice, *adj.*, silly, small, trifling

Nicely, *adv.*, sillily

Nother, *con.*, neither

Noyful, *adj.*, noxious, hurtful

Noying, *n.*, harm


Ordinate, *adj.*, ordered, prescribed

Otherdeal, *adv.*, otherwise

Overthwart, *adj.*, crossed over on itself

Passing, *adj.*, surpassing

Patent, *n.*, a plate or paten (patine)

Pight, *adj.*, put, pitched

Powder, *n.*, dust of any kind

Pricket, *n.*, a spike used for candlestick, hence a candle

Principles, *n.*, indecomposable elements

Pure, *v.a.*, to purify

Pursueth, *v*, suiteth?


Quicken, *v.i.*, to come to life

Quiver, *adj.*, nimble, active


Ramaious, *adj.*, (hawking), slow

Ravish, *v.*, to snatch

Reclaim, *n.*, (hawking}, the calling back of a hawk

Refudation, *n.*, a process in which vinegar is poured on lead, distilled off, and again suffered to act on it

Relief, *n.*, a dessert

Rese, *v.*, to rush on anyone

Resolve, *v.*, to loosen, weaken, to dissolve

Rheum, *n.*, salt humour

Ribbed, *adj.*, beaten with a "rib," in dressing flax

Ridge, *n.*, the back bone

Riever, *n.*, a violent, robber, a raider

Rivelled, *adj.*, wrinkled

Rively, *adv.*, wrinkled, shrunk

Rodded, *adj.*, separated from tow--"redded"

Routs, *n.*, crowds

Ruthful, *adj.*, sorrowful


Sad, *adj.*, steadfast, solid

Sanguine, *adj.*, blood-like

Scarce, *adj.*, sparing, avaricious

Seethe, *v.*, to boil

Selde, *adv.*, seldom

Sele, *v.*, to cover

Shamefast, *adj.*, shamefaced

Sheltrons, *n.*, palisades

Shern, *adj.*, shore

Shewer, *n.*, a looking-glass

Shingle, *n.*, in *roofing*, brushwood, or small boards

Shipbreach, *n.*, shipwreck

Shore, *adj.*, shorn (of the hair)

Shrewd, *adj.*, bitter; *cf.*, shrew

Silly, *adj.*, blessed, *hence* innocent, *hence* simple

Sinew, *n.*, a nerve

Slubber, *v.*, to do anything carelessly

Smirch, *v.*, to soil

Sod, *adj.*, stewed

Solar, *n.*, an upper floor

Solemn, *adj.*, celebrated, earnest

Somedeal, *adv.*, somewhat

Sometime, *adv.*, once

Sooth, *n.*, truth

Spanells, *n.*, ropes connecting the fore or hind feet of an animal to impede its movements

Spousehood, *n.*, marriage

Spousebreach, *n.*, adultery

Spronge, *adj.*, sprinkled

Stare, *v.*, to stay

Startling, *part.*, leaping and jumping
Stint, *v.*, to stop
Stint, *adj.*, stopped
Straight, *adj.*, confined
Straited, *adj.*, narrowed
Sty, *v.*, to climb
Suspect, *adj.*, in suspicion

Tatch, *n.*, spot
Tatched, *adj.*, spotted
Tewly, livid
Tilth, *v.*, to cultivate
Tilth, *n.*, tillage
Tofore, *prep.*, before
Toll, *v.*, to entice
Trow, *v.*, to believe; *cf.*, trust

Unmighty, *adj.*, unable
Unneth, *adv.*, hardly
Uplandish, *adj.*, rustic
Utter, *adj.*, outer

Very, *adj.*, true

Wait, *n.*, a guard
Wanhope, *n.*, despair
Warily, *adv.*, carefully
Ween, *v.*, consider, think
Wem, *n.*, blemish, fault
Wend, *adj.*, wound up
Werish, *adj.*, insipid

Whelk, *n.*, a swelling

Whet, *v.*, to sharpen

Whiche, *n.*, a wicket-gate *cf.*, "wych gate"

Wilful, *adj.*, of set purpose

Wit, *n.*, a sense; *cf* "out of his wits"

Witty, *adj.*, sensibly

Wonder, *adj.*, wondrous

Wonderly, *adv.*, wondrously

Wood, *adj.*, crazy, frantic

Woodness, *n.*, madness

Woose, *n.*, fluid

Worship, *n.*, reverence, authority

Wosen, *n.*, the arteries

Wot, *v.*, knew

Wrang, *adj.*, injured, wrung

Wreak, *n.*, revenge

Wreck, *v.*, to revenge

Wrecker, *n.*, avenger

www.bookjungle.com *email: sales@bookjungle.com fax: 630-214-0564 mail: Book Jungle PO Box 2226 Champaign, IL 61825*

## The Codes Of Hammurabi And Moses
### W. W. Davies

QTY

The discovery of the Hammurabi Code is one of the greatest achievements of archaeology, and is of paramount interest, not only to the student of the Bible, but also to all those interested in ancient history...

**Religion**     **ISBN:** *1-59462-338-4*     **Pages:132**

*MSRP $12.95*

## The Theory of Moral Sentiments
### Adam Smith

QTY

This work from 1749. contains original theories of conscience amd moral judgment and it is the foundation for systemof morals.

**Philosophy**   **ISBN:** *1-59462-777-0*     **Pages:536**

*MSRP $19.95*

## Jessica's First Prayer
### Hesba Stretton

QTY

In a screened and secluded corner of one of the many railway-bridges which span the streets of London there could be seen a few years ago, from five o'clock every morning until half past eight, a tidily set-out coffee-stall, consisting of a trestle and board, upon which stood two large tin cans, with a small fire of charcoal burning under each so as to keep the coffee boiling during the early hours of the morning when the work-people were thronging into the city on their way to their daily toil...

**Pages:84**

**Childrens**    **ISBN:** *1-59462-373-2*     *MSRP $9.95*

## My Life and Work
### Henry Ford

QTY

Henry Ford revolutionized the world with his implementation of mass production for the Model T automobile. Gain valuable business insight into his life and work with his own auto-biography... "We have only started on our development of our country we have not as yet, with all our talk of wonderful progress, done more than scratch the surface. The progress has been wonderful enough but..."

**Pages:300**

**Biographies/**   **ISBN:** *1-59462-198-5*     *MSRP $21.95*

## The Art of Cross-Examination
## Francis Wellman

QTY

I presume it is the experience of every author, after his first book is published upon an important subject, to be almost overwhelmed with a wealth of ideas and illustrations which could readily have been included in his book, and which to his own mind, at least, seem to make a second edition inevitable. Such certainly was the case with me; and when the first edition had reached its sixth impression in five months, I rejoiced to learn that it seemed to my publishers that the book had met with a sufficiently favorable reception to justify a second and considerably enlarged edition. ..

**Pages:412**

Reference    ISBN: *1-59462-647-2*    *MSRP $19.95*

## On the Duty of Civil Disobedience
## Henry David Thoreau

QTY

Thoreau wrote his famous essay, On the Duty of Civil Disobedience, as a protest against an unjust but popular war and the immoral but popular institution of slave-owning. He did more than write—he declined to pay his taxes, and was hauled off to gaol in consequence. Who can say how much this refusal of his hastened the end of the war and of slavery ?

Law          ISBN: *1-59462-747-9*          **Pages:48**

*MSRP $7.45*

## Dream Psychology Psychoanalysis for Beginners
## Sigmund Freud

QTY

Sigmund Freud, born Sigismund Schlomo Freud (May 6, 1856 - September 23, 1939), was a Jewish-Austrian neurologist and psychiatrist who co-founded the psychoanalytic school of psychology. Freud is best known for his theories of the unconscious mind, especially involving the mechanism of repression; his redefinition of sexual desire as mobile and directed towards a wide variety of objects; and his therapeutic techniques, especially his understanding of transference in the therapeutic relationship and the presumed value of dreams as sources of insight into unconscious desires.

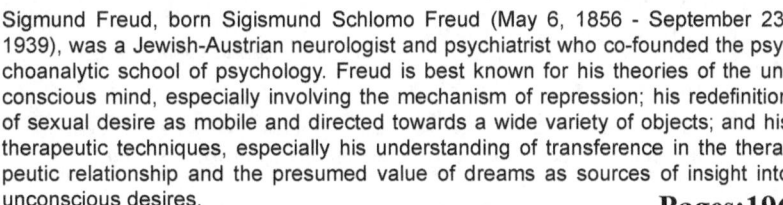

**Pages:196**

Psychology    ISBN: *1-59462-905-6*    *MSRP $15.45*

## The Miracle of Right Thought
## Orison Swett Marden

QTY

Believe with all of your heart that you will do what you were made to do. When the mind has once formed the habit of holding cheerful, happy, prosperous pictures, it will not be easy to form the opposite habit.  It does not matter how improbable or how far away this realization may see, or how dark the prospects may be, if we visualize them as best we can, as vividly as possible, hold tenaciously to them and vigorously struggle to attain them, they will gradually become actualized, realized in the life. But a desire, a longing without endeavor, a yearning abandoned or held indifferently will vanish without realization.

**Pages:360**

Self Help        ISBN: *1-59462-644-8*    *MSRP $25.45*

**The Rosicrucian Cosmo-Conception Mystic Christianity** *by Max Heindel*  ISBN: *1-59462-188-8*  **$38.95**
*The Rosicrucian Cosmo-conception is not dogmatic, neither does it appeal to any other authority than the reason of the student. It is: not controversial, but is: sent forth in the, hope that it may help to clear...*  New Age/Religion Pages 646

**Abandonment To Divine Providence** *by Jean-Pierre de Caussade*  ISBN: *1-59462-228-0*  **$25.95**
*"The Rev. Jean Pierre de Caussade was one of the most remarkable spiritual writers of the Society of Jesus in France in the 18th Century. His death took place at Toulouse in 1751. His works have gone through many editions and have been republished...*  Inspirational/Religion Pages 400

**Mental Chemistry** *by Charles Haanel*  ISBN: *1-59462-192-6*  **$23.95**
*Mental Chemistry allows the change of material conditions by combining and appropriately utilizing the power of the mind. Much like applied chemistry creates something new and unique out of careful combinations of chemicals the mastery of mental chemistry...*  New Age Pages 354

**The Letters of Robert Browning and Elizabeth Barret Barrett 1845-1846 vol II**  ISBN: *1-59462-193-4*  **$35.95**
*by Robert Browning and Elizabeth Barrett*  Biographies Pages 596

**Gleanings In Genesis (volume I)** *by Arthur W. Pink*  ISBN: *1-59462-130-6*  **$27.45**
*Appropriately has Genesis been termed "the seed plot of the Bible" for in it we have, in germ form, almost all of the great doctrines which are afterwards fully developed in the books of Scripture which follow...*  Religion/Inspirational Pages 420

**The Master Key** *by L. W. de Laurence*  ISBN: *1-59462-001-6*  **$30.95**
*In no branch of human knowledge has there been a more lively increase of the spirit of research during the past few years than in the study of Psychology, Concentration and Mental Discipline. The requests for authentic lessons in Thought Control, Mental Discipline and...*  New Age/Business Pages 422

**The Lesser Key Of Solomon Goetia** *by L. W. de Laurence*  ISBN: *1-59462-092-X*  **$9.95**
*This translation of the first book of the "Lemegton" which is now for the first time made accessible to students of Talismanic Magic was done, after careful collation and edition, from numerous Ancient Manuscripts in Hebrew, Latin, and French...*  New Age/Occult Pages 92

**Rubaiyat Of Omar Khayyam** *by Edward Fitzgerald*  ISBN:*1-59462-332-5*  **$13.95**
*Edward Fitzgerald, whom the world has already learned, in spite of his own efforts to remain within the shadow of anonymity, to look upon as one of the rarest poets of the century, was born at Bredfield, in Suffolk, on the 31st of March, 1809. He was the third son of John Purcell...*  Music Pages 172

**Ancient Law** *by Henry Maine*  ISBN: *1-59462-128-4*  **$29.95**
*The chief object of the following pages is to indicate some of the earliest ideas of mankind, as they are reflected in Ancient Law, and to point out the relation of those ideas to modern thought.*  Religiom/History Pages 452

**Far-Away Stories** *by William J. Locke*  ISBN: *1-59462-129-2*  **$19.45**
*"Good wine needs no bush, but a collection of mixed vintages does. And this book is just such a collection. Some of the stories I do not want to remain buried for ever in the museum files of dead magazine-numbers an author's not unpardonable vanity..."*  Fiction Pages 272

**Life of David Crockett** *by David Crockett*  ISBN: *1-59462-250-7*  **$27.45**
*"Colonel David Crockett was one of the most remarkable men of the times in which he lived. Born in humble life, but gifted with a strong will, an indomitable courage, and unremitting perseverance...*  Biographies/New Age Pages 424

**Lip-Reading** *by Edward Nitchie*  ISBN: *1-59462-206-X*  **$25.95**
*Edward B. Nitchie, founder of the New York School for the Hard of Hearing, now the Nitchie School of Lip-Reading, Inc, wrote "LIP-READING Principles and Practice". The development and perfecting of this meritorious work on lip-reading was an undertaking...*  How-to Pages 400

**A Handbook of Suggestive Therapeutics, Applied Hypnotism, Psychic Science**  ISBN: *1-59462-214-0*  **$24.95**
*by Henry Munro*  Health/New Age/Health/Self-help Pages 376

**A Doll's House: and Two Other Plays** *by Henrik Ibsen*  ISBN: *1-59462-112-8*  **$19.95**
*Henrik Ibsen created this classic when in revolutionary 1848 Rome. Introducing some striking concepts in playwriting for the realist genre, this play has been studied the world over.*  Fiction/Classics/Plays 308

**The Light of Asia** *by sir Edwin Arnold*  ISBN: *1-59462-204-3*  **$13.95**
*In this poetic masterpiece, Edwin Arnold describes the life and teachings of Buddha. The man who was to become known as Buddha to the world was born as Prince Gautama of India but he rejected the worldly riches and abandoned the reigns of power when...*  Religion/History/Biographies Pages 170

**The Complete Works of Guy de Maupassant** *by Guy de Maupassant*  ISBN: *1-59462-157-8*  **$16.95**
*"For days and days, nights and nights, I had dreamed of that first kiss which was to consecrate our engagement, and I knew not on what spot I should put my lips..."*  Fiction/Classics Pages 240

**The Art of Cross-Examination** *by Francis L. Wellman*  ISBN: *1-59462-309-0*  **$26.95**
*Written by a renowned trial lawyer, Wellman imparts his experience and uses case studies to explain how to use psychology to extract desired information through questioning.*  How-to/Science/Reference Pages 408

**Answered or Unanswered?** *by Louisa Vaughan*  ISBN: *1-59462-248-5*  **$10.95**
*Miracles of Faith in China*  Religion Pages 112

**The Edinburgh Lectures on Mental Science (1909)** *by Thomas*  ISBN: *1-59462-008-3*  **$11.95**
*This book contains the substance of a course of lectures recently given by the writer in the Queen Street Hall, Edinburgh. Its purpose is to indicate the Natural Principles governing the relation between Mental Action and Material Conditions...*  New Age/Psychology Pages 148

**Ayesha** *by H. Rider Haggard*  ISBN: *1-59462-301-5*  **$24.95**
*Verily and indeed it is the unexpected that happens! Probably if there was one person upon the earth from whom the Editor of this, and of a certain previous history, did not expect to hear again...*  Classics Pages 380

**Ayala's Angel** *by Anthony Trollope*  ISBN: *1-59462-352-X*  **$29.95**
*The two girls were both pretty, but Lucy who was twenty-one who supposed to be simple and comparatively unattractive, whereas Ayala was credited, as her Bombwhat romantic name might show, with poetic charm and a taste for romance. Ayala when her father died was nineteen...*  Fiction Pages 484

**The American Commonwealth** *by James Bryce*  ISBN: *1-59462-286-8*  **$34.45**
*An interpretation of American democratic political theory. It examines political mechanics and society from the perspective of Scotsman James Bryce*  Politics Pages 572

**Stories of the Pilgrims** *by Margaret P. Pumphrey*  ISBN: *1-59462-116-0*  **$17.95**
*This book explores pilgrims religious oppression in England as well as their escape to Holland and eventual crossing to America on the Mayflower, and their early days in New England...*  History Pages 268

QTY

**The Fasting Cure** by *Sinclair Upton*  ISBN: *1-59462-222-1*  **$13.95**
*In the Cosmopolitan Magazine for May, 1910, and in the Contemporary Review (London) for April, 1910, I published an article dealing with my experiences in fasting. I have written a great many magazine articles, but never one which attracted so much attention...  New Age/Self Help/Health Pages 164*

**Hebrew Astrology** by *Sepharial*  ISBN: *1-59462-308-2*  **$13.45**
*In these days of advanced thinking it is a matter of common observation that we have left many of the old landmarks behind and that we are now pressing forward to greater heights and to a wider horizon than that which represented the mind-content of our progenitors...  Astrology Pages 144*

**Thought Vibration or The Law of Attraction in the Thought World**  ISBN: *1-59462-127-6*  **$12.95**
by *William Walker Atkinson*  *Psychology/Religion Pages 144*

**Optimism** by *Helen Keller*  ISBN: *1-59462-108-X*  **$15.95**
*Helen Keller was blind, deaf, and mute since 19 months old, yet famously learned how to overcome these handicaps, communicate with the world, and spread her lectures promoting optimism. An inspiring read for everyone...  Biographies/Inspirational Pages 84*

**Sara Crewe** by *Frances Burnett*  ISBN: *1-59462-360-0*  **$9.45**
*In the first place, Miss Minchin lived in London. Her home was a large, dull, tall one, in a large, dull square, where all the houses were alike, and all the sparrows were alike, and where all the door-knockers made the same heavy sound...  Childrens/Classic Pages 88*

**The Autobiography of Benjamin Franklin** by *Benjamin Franklin*  ISBN: *1-59462-135-7*  **$24.95**
*The Autobiography of Benjamin Franklin has probably been more extensively read than any other American historical work, and no other book of its kind has had such ups and downs of fortune. Franklin lived for many years in England, where he was agent...  Biographies/History Pages 332*

| | |
|---|---|
| **Name** | |
| **Email** | |
| **Telephone** | |
| **Address** | |
| | |
| **City, State ZIP** | |

☐ **Credit Card**          ☐ **Check / Money Order**

| | |
|---|---|
| **Credit Card Number** | |
| **Expiration Date** | |
| **Signature** | |

*Please Mail to:  Book Jungle*
*PO Box 2226*
*Champaign, IL 61825*
*or Fax to:        630-214-0564*

## ORDERING INFORMATION

**web***: www.bookjungle.com*
**email***: sales@bookjungle.com*
**fax***: 630-214-0564*
**mail***: Book Jungle  PO Box 2226  Champaign, IL 61825*
**or PayPal** *to sales@bookjungle.com*

***Please contact us for bulk discounts***

## DIRECT-ORDER TERMS

**20% Discount if You Order
Two or More Books**
Free Domestic Shipping!
Accepted: Master Card, Visa,
Discover, American Express